# AN ATLAS OF SELECTED GALAXIES

# AN ATLAS OF SELECTED GALAXIES

## with Illustrations of Photometric Analyses


Edited by

Bunshiro Takase
Tokyo Astronomical Observatory, University of Tokyo

Keiichi Kodaira
Tokyo Astronomical Observatory, University of Tokyo

Sadanori Okamura
Tokyo Astronomical Observatory, University of Tokyo



UNIVERSITY OF TOKYO PRESS
VNU SCIENCE PRESS

Printed in Japan



Published jointly by:
UNIVERSITY OF TOKYO PRESS, Tokyo
ISBN 4-13-068103-6 (UTP 69030)

and

VNU SCIENCE PRESS BV, Utrecht The Netherlands
ISBN 90-6764-007-7

Distributed in all areas outside Asia by VNU Science Press BV, Utrecht The Netherlands

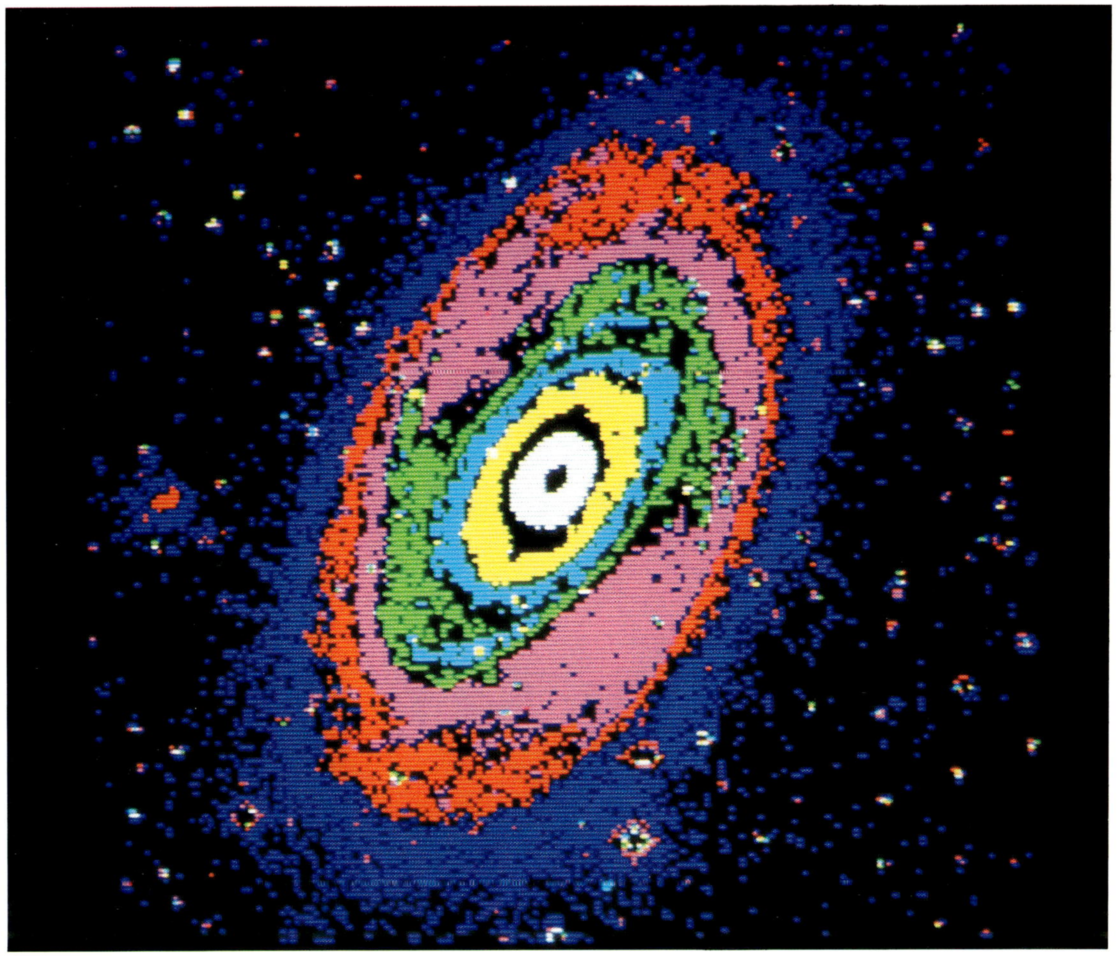

(a)

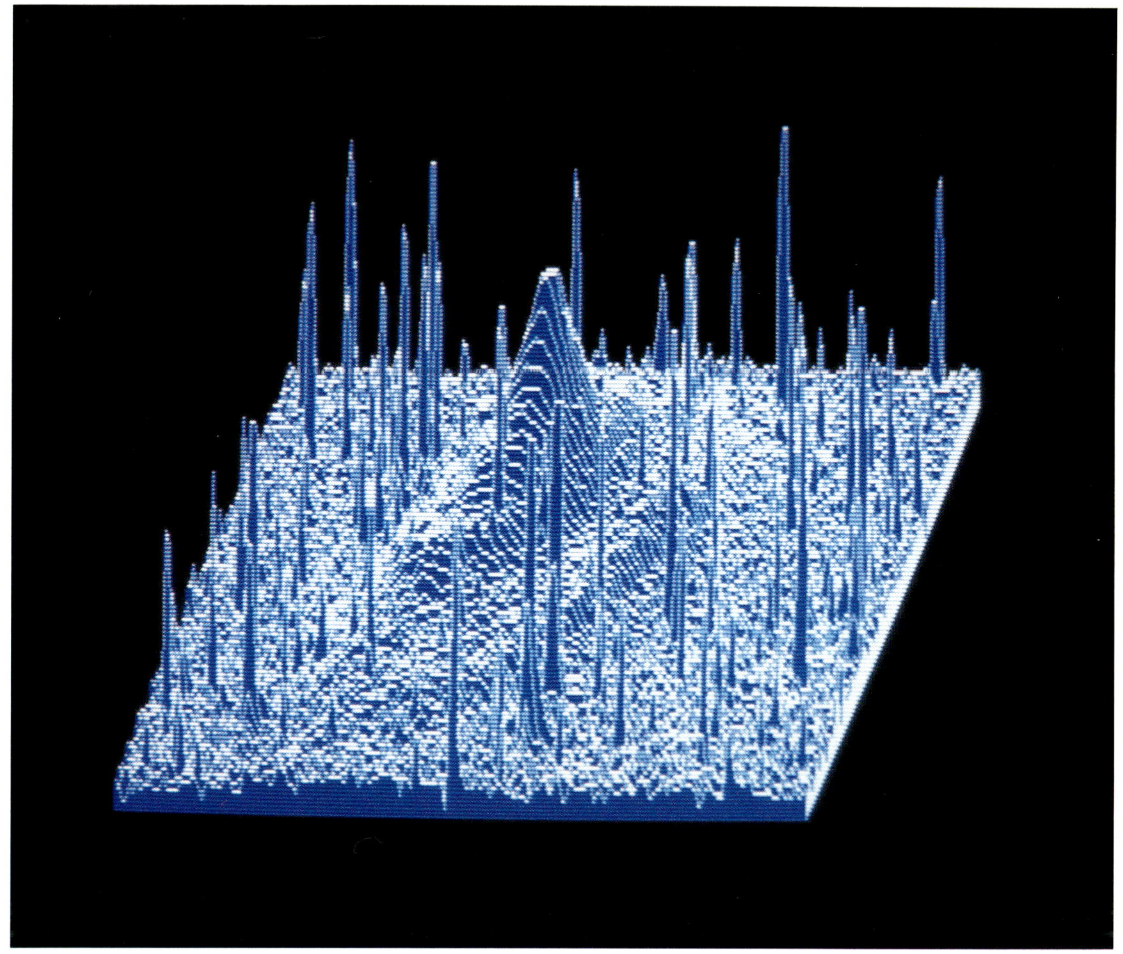

(b)

In addition to its many photographs, this atlas illustrates the computerized analysis of galaxy photographs. Presented here are examples of images constructed by a computer using various digital image processing techniques.

(a) A false-color image of NGC 3031 (M81) showing the details of its luminosity distribution.

(b) Three-dimensional representation of the luminosity distribution of NGC 3031(M81).

A0.1: NGC 224(M31) SA(s)b (B) ⓘ

A0.2a: NGC 598 (M33) SA(s)cd (V)

A0.2b: NGC 598 (M33) SA(S)cd (U)

A0.3a: NGC 5194/5195 (M51) SA(s)bcP/I0P (B)

A0.3b: NGC 5194/5195 (M51) SA(s)bcP/I0P (I)

# Table of Contents

# Preface

It is only several decades since galaxies were recognized as huge aggregations of matter in the universe which consist of myriads of stars and vast clouds of gas. The wide variety of shapes of galaxies is one of the wonders of the universe. Most people enjoy looking at photographs of galaxies and often feel curious about why they look so diverse.

Attempts to classify galaxies have continued since an early stage of study, when they were still called 'nebulae'. As in some other fields of science, the first step of classification was a grouping of the patterns, and the second step a systematic arrangement of various morphological types. Several classification schemes currently used, such as Hubble's, are in this second stage. These classifications are usually based on pattern recognition in a visual inspection of the photographic image of a galaxy, and therefore essentially qualitative and inevitably more or less subjective.

The next step, then, should be to proceed toward a quantitative and objective classification. Such an approach may be best achieved through the surface photometry of galaxies, a technique which has advanced remarkably in recent years. Photographs of galaxies taken in several color bands accumulated over years can now be efficiently reduced to photometric data with the help of high-performance measuring instruments and computerized image processing techniques. By analyzing these data in detail, quantitative information on the luminosity and color distributions for every element of galaxies may be obtained.

This atlas presents the photographs of selected galaxies taken with either the Okayama 188-cm reflector (since 1960) or the Kiso 105-cm Schmidt telescope (since 1975) for the purpose of photometric analyses. A number of illustrations are given to demonstrate how photographs of galaxies are reduced into quantitative photometric data. These illustrations also show some representative results obtained in the course of our work on the quantitative classification of galaxies.

We wish to acknowledge the assistance given to our observations by the staffs of the Okayama Astrophysical Observatory and the Kiso Observatory (both are branches of the Tokyo Astronomical Observatory) and of several graduate students of the Department of Astronomy, University of Tokyo. Photographs were prepared using the photographic facilities of the Kiso Observatory and the Department of Astronomy, University of Tokyo. Computations for our photometric works were carried out at the Computer Centre, University of Tokyo, and in part using the computing facilities of the Tokyo Astronomical Observatory.

We acknowledge the financial support, primarily in the form of Research Fund of the Ministry of Education, Science and Culture, provided at various stages of the photometric works. We also wish to thank many friends who gave us continuous encouragement over the long period of our research. Special thanks go to Mr. C. R. Morgan who kindly improved the manuscript, and to Dr. K. Wakamatsu who made his plates available to us.

June 1983

Bunshiro Takase
Keiichi Kodaira
Sadanori Okamura

# Introduction

In the interest of effective utilization of this atlas, a general description of the classification systems and photometric analyses of galaxies, together with some results obtained by our research team, are given in Part I. This part may serve as an explanation of the atlas and illustrations presented in Part II.

Part I includes eight sections written by the authors listed in the table of contents, who all shared the photometric work included in this atlas and prepared the illustrations in collaboration with the editors. Sections 1 and 2 give brief reviews of the classification of galaxies and the concept of surface photometry, respectively. Section 3 summarizes the standard method of analysis of the radial luminosity distribution. Section 4 explains the method of analysis of the luminosity distribution perpendicular to the galactic plane for edge-on galaxies. In section 5 a newly introduced device, the 'projected profile', and related analyses are presented. Section 6 describes a method of spectral analysis for the quantitative representation of spiral patterns and other substructures. Finally, sections 7 and 8 present explanations of active and peculiar galaxies and clusters of galaxies.

The references are given at the end of Part I. They are divided into three categories: general references (G), our publications related to this atlas (P), and representative atlases and catalogues (C).

Apart from several frontispieces, all the illustrations in Part II are arranged in order of sections 1 to 8 in Part I. Each illustration can be identified by a figure number at the bottom. Figures in Part II are headed by the character A, which denotes the atlas, followed by the section number and the sequential number in the section. For example, figure A3.1b means figure 1b in section 3 of Part II; figure A0.1a, figure 1a in the frontispieces. For each photograph, the following items are given: (1) figure number, (2) identification of the object, (3) morphological type in RC 2 catalogue (C5) for galaxies or representative types for clusters of galaxies (G7; G53), (4) color band, (5) mark indicating the north direction and (6) other relevant data from photometric analysis, if any.

Part III includes tables of data for the atlas. All the galaxies and the clusters of galaxies included in this atlas are summarized in table 1. The list of the original plates is given in table 2, and detailed data on the isophotal maps are given in table 3.

# Part I. General Descriptions of the Classification and Photometric Analyses

# 1. Classification of Galaxies

## 1.1. Classifications in Early Period

In 1786, William Herschel (G21) published the first volume of his catalogue of nebulae and star clusters, classifying them into five types of nebulae — bright, faint, very faint, planetary, and very large — and three types of clusters. His son, John Herschel (G22) compiled 'A General Catalogue of Nebulae and Clusters of Stars', and published it in 1864. He introduced a classification system which employed five physical attributes — magnitude, brightness, roundness, condensation, and resolvability — each with five subdivisions.

In 1908, Wolf (G68) presented 23 drawings of nebulae as standards. Then in 1919, Curtis (G9) grouped nebulae into three types — planetaries, diffuse, and spirals. Hubble (G24) proposed the following scheme of classification as 'a compromise between Curtis' generalization and Wolf's specialization':

Galactic nebulae: Planetary and Diffuse (luminous and dark)
Non-galactic nebulae: Sprial, Spindle, Ovate, Globular and Irregular

Here 'non-galactic' means that these objects are not concentrated in the galactic plane.

Around that time Jeans (G28) had developed an evolutionary theory of spiral nebulae in his book 'Problems of Cosmogony and Stellar Dynamics'. Hubble tried to construct a classification scheme analogous to Jeans' evolutionary sequence but, according to him, based purely on observational material. He suggested the following grouping for the non-galactic nebulae in 1923 (G25):

Amorphous: A0, A1, A2 and A3
Spirals: S0, S1, S2 and S3
Others: $\phi$-shaped, S-shaped and Irregulars

A0 denotes round amorphous nebulae, and in order of A1, A2 and A3 the flatness or ellipticity increases. Meanwhile as one goes from S0 to S3, the spiral arm becomes less tightly wound.

## 1.2. Hubble's Classification

In 1926 Hubble (G26) published a new classification system of extra-galactic nebulae, extending his previous scheme as follows:

A. Regular
1. Elliptical: E*n* (*n* = 0,1,...., 7. The numeral *n* indicates the ellipticity multiplied by 10.)
2. Spirals
   a) Normal spirals: Sa (early), Sb (intermediate) and Sc (late)
   b) Barred spirals: SBa (early), SBb (intermediate) and SBc (late)

B. Irregular: Irr

The term 'early' or 'late' implies the sequential order of forms, which continues from the elliptical series. As one proceeds from Sa (SBa) to Sc (SBc), (a) the size of the central bulge decreases, (b) the spiral arms become less tightly wound and (c) the degree of resolution into stars increases.

In his book 'The Realm of the Nebulae' published in 1936, Hubble (G27) presented the so-called tuning fork diagram (Fig. 1.1); at the junction of the fork he reintroduced the S0 type from his 1923 scheme, which he described as follows: 'The junction may be represented by

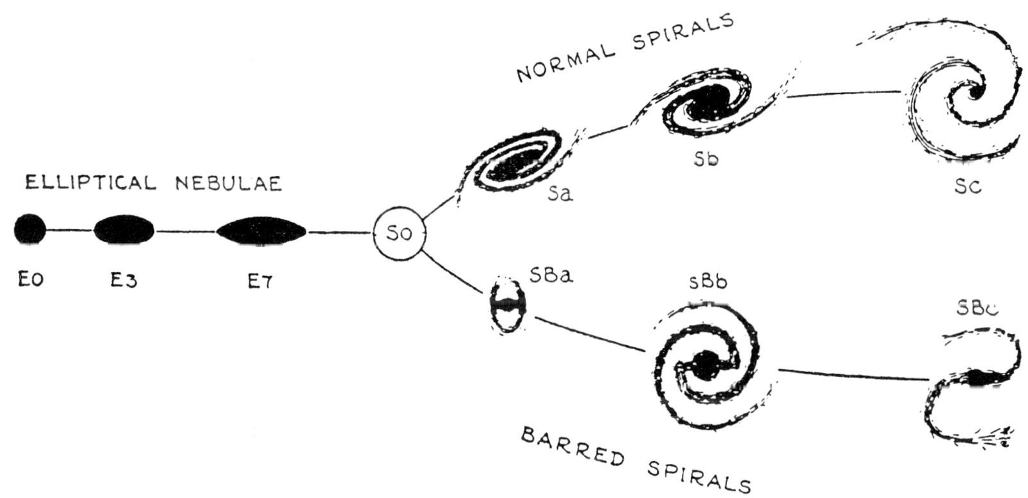


Fig. 1.1. Hubble's tuning fork diagram of the nebular types.

the more or less hyphothetical class S0 — a very important stage in all theories of nebular evolution. Observations suggest a smooth transition between E7 and SBa, but indicate a dicontinuity between E7 and Sa in the sense that Sa spirals are always found with arms fully developed. Speculation concerning the discontinuity will be unprofitable until more detailed information has been assembled from large-scale photographs.'

Hubble later succeeded in discovering many S0 galaxies which were flatter than E7 and had neither spiral arms nor bars, in the photographs of nearly 1000 bright galaxies taken with 60-inch and 100-inch telescopes at Mt. Wilson. Then, toward 1950, he modified his tuning fork diagram into the following form, which is described in 'The Hubble Atlas of Galaxies' published by Sandage (C2) in 1961:

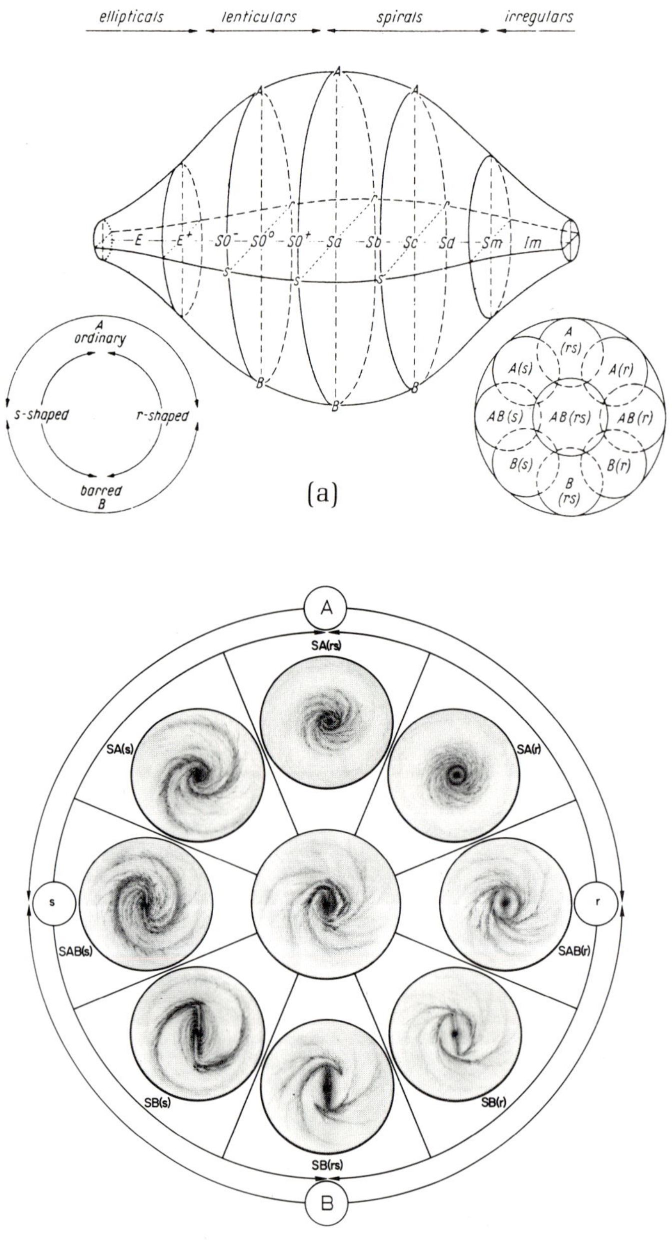


(b)

Fig. 1.2. The classification volume of de Vaucouleurs and its plane projections (a), and one of its cross-section near the region of the Sb spiral (b). (Courtesy of Dr. G. de Vaucouleurs).

S0 — Sa — Sb — Sc

E0 — E1 — ...... — E7

SB0 — SBa — SBb — SBc

Here S0 has changed its position, while SB0 corresponds to SBa of Hubble's former sequence. The galaxies previously classified as SBb are divided into SBa and SBb in this new scheme.

Irregular galaxies are grouped into two types. The first one, denoted by Irr I, includes those with highly resolved O and B type stars and H II regions, while the second, Irr II, includes those whose images are completely smooth in texture.

## 1.3. De Vaucouleurs' Classification (Revised Hubble's Classification)

De Vaucouleurs (G11) developed a revised system of galaxy classification in 1956 when he published a list of southern galaxies which he observed at Mt. Stromlo Observatory in Australia. His classification includes the additional type and varieties to Hubble's system:

A) Classes: E (elliptical), S0 (lenticular), S (spiral) and I (irregular)
B) Families of the lenticulars and spirals:
A (normal), B (barred) and AB (intermediate)
C) Stages of the spirals: a (early), b (intermediate), c (late), d (very late) and m (Magellanic)
C') Stages of the lenticulars:
$S0^-$ (early), $S0^0$ (intermediate) and $S0^+$ (late)
D) Varieties of the lenticulars and spirals:
r (ringed type), s (s-shaped) and rs (intermediate)

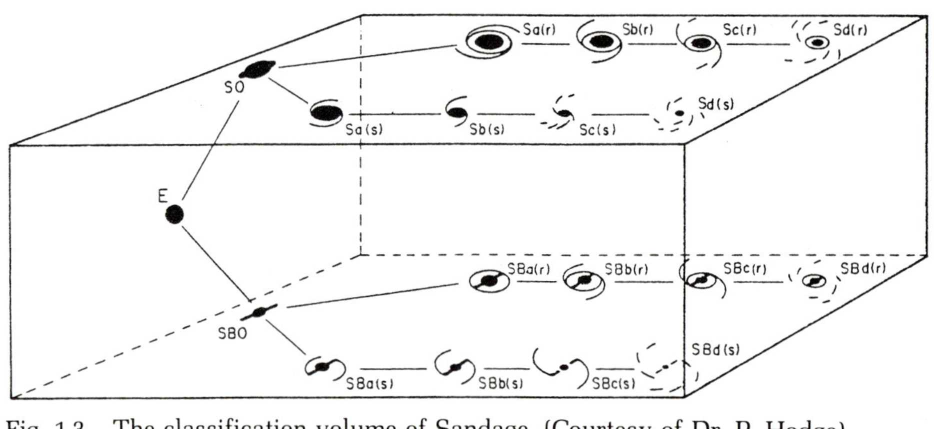


Fig. 1.3. The classification volume of Sandage. (Courtesy of Dr. P. Hodge).

For irregulars de Vaucouleurs gives two groups, namely,

Im: Magellanic irregulars and
I0: Non-Magellanic irregulars which have structure related to the S0 class.

Figure 1.2 gives a three-dimensional representation of his classification scheme, which he calls the classification volume, and its cross-section. He and his collaborators have been continuing to classify a large number of galaxies according to this scheme (G8; G12; G14).

In the above-mentioned 'Hubble Atlas of Galaxies' Sandage also presents a classification volume, further modifying Hubble's scheme by introducing de Vaucouleurs' designation of the varieties. Figure 1.3 is an illustration of this volume originally presented by Hodge (G23). Sandage and Tammann (C6) recently completed a reclassification of 1246 bright galaxies originally compiled by Shapley and Ames (G50) on the basis of this scheme.

## 1.4. Van den Bergh's Luminosity Class (DDO Classification)

In 1960, when he was at the David Dunlap Observatory, van den Bergh (G59–G61) showed that a correlation exists between the appearance of the spiral arms and the absolute luminosity of galaxies. It has been revealed that the more luminous spiral and irregular galaxies have longer, more developed arms and/or bars with higher surface brightness. By inspecting galaxy images on the Palomar Sky Survey prints, he classified some 500 Sb and Sc and about 20 irregular galaxies into five luminosity classes, according to the degree of development of their arms and bars.

The luminosity class designation, which is sometimes called the DDO classification, is as follows:

I (supergiant), II (bright giant), III (normal giant), IV (subgiant) and V (dwarf)

Van den Bergh's calibration of the photographic absolute magnitude for each class, $M_{pg}$, is given in the list below:

| Type: | SbI | SbII | SbIII | ScI | Sc&IrrII | Sc&IrrIII | Sc&IrrIV | Sc&IrrV |
|---|---|---|---|---|---|---|---|---|
| $M_{pg}$: | -20.4 | -19.4 | -18.0 | -20.0 | -19.4 | -18.3 | -17.3: | -15: |

Recent work on the luminosity class by Tammann et al. (G55), however, revealed a larger spread in the absolute magnitude within each luminosity class than expected earlier.

## 1.5. Morgan's Classification (Yerkes Classification)

From spectroscopic work carried out by Morgan and Mayall (G40), it has been revealed that a relation exists between the composite spectra of galaxies and the degree of dominance of their spheroidal component. Spectra in the violet wavelength region of irregulars and late-type spirals having little or no central luminosity concentration indicate that their luminosity is due principally to B, A and F type stars, while those of giant ellipticals and spirals with extreme central luminosity concentrations show that their light is mainly contributed by K type stars.

With this information Morgan (G38; G39) introduced a new classification scheme in 1958 and 1959, which is primarily based on the degree of luminosity concentration or degree of dominance of the central spheroidal component of the galaxy. Morgan's system, which is also known as the Yerkes classification, consists of the following items.

A) Fundamental parameter (degree of luminosity concentration):
a (lowest), f (low), g (higher) and k (highest)

According to Morgan, galaxies with the designations a, f, g or k show composite spectra in violet region approximately equivalent to those of A-, F-, G- or K-type stars, respectively, although the designations are determined entirely from the degree of central luminosity.

B) Form family:
S (spirals), B (barred spirals), E (ellipticals), I (irregulars), Ep (elliptical systems containing dust absorption or other peculiarities), D (systems showing rotational sysmmetry without pronounced spiral or elliptical structure), L (systems of low surface brightness) and N (systems having small, brilliant nuclei superposed on a considerably fainter background)

C) Inclination class:
Not only for ellipticals but also for all other types, a numeral designation 1, 2, ... , 7 is given to indicate their apparent ellipticity.

Figures A1.1 through A1.29 show various morphological types of galaxies with classification designations.

## 1.6. Remarks on the New Step of Galaxy Classification

In chapter 1 of the book 'Galaxies and the Universe', Sandage (G46) mentions that a simple description of morphological features is the first step in the classification of an object, and that Wolf's (G68) and Vorontsov-Velyaminov's (C7–C10) classifications of galaxies are examples of this type. Further, he says that the second step is to group the objects of each set into classes according to some continuously varying parameter. The classification schemes of Hubble, de Vaucouleurs, van den Bergh and Morgan are all included in this category.

Whichever scheme is used, the classes of galaxies have been determined mainly by visual inspection of the photographic image, based on morphological features such as the general shape, the appearance of spiral arms and bars, the relative proportion of the spheroidal component in size and luminosity, and so on. Galaxy classifications conducted in this way are essentially qualitative and inevitably subjective to some degree.

The next step, then, should be to proceed toward a quantitative classification. Recently the techniques of surface photometry of galaxies have advanced dramatically. Luminosity distributions in various color bands can be accurately obtained and analyzed using computer-based methods. The qualitative representation of galactic features from visual estimation should now be replaced by quantitative indicators which are more objective since they are free from personal differences of judgement.

Several attempts at this kind of quantitative classification have been in progress here and there. Our group is also making an effort to establish a new quantitative classification system. The results achieved so far by our approach are described briefly in the appendix.

# 2. Surface Photometry

## 2.1. Quantitative Presentation of Photographic Images

All of the classical classifications of galaxies are based upon visual inspections of morphology of galaxies in photographs. A method for the quantitative evaluation of photographs became available with the introduction of the technique of surface photometry, in which the darkness or density of negative photographic plates are calibrated to the flux of radiation to which the plates were exposed. This calibration can be performed by exposing a part of the photographic plate to uniform light through step wedges with varying but known relative transmittance. Thus one can convert the density of negative photographs into surface brightness, which is conventionally expressed in the unit of magnitude per square second of arc (mag arcsec$^{-2}$). The zero point of the magnitude scale is determined with the help of some objects in the same photographic field whose brightnesses are known.

In order to obtain the surface brightness at each point of a galaxy, one has to measure

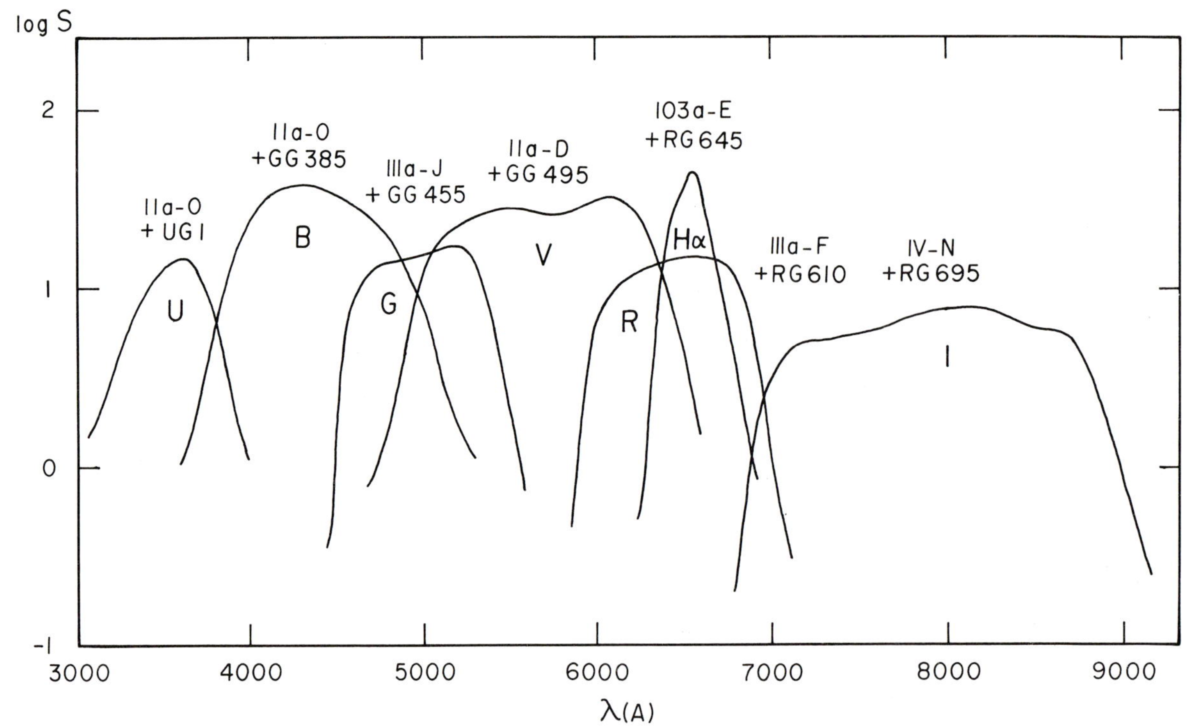


Fig. 2.1. Relative sensitivity curves of the color bands, produced by the combination of the Kodak spectroscopic emulsion and the Schott glass filter indicated at the top of each band.

carefully the density of each picture element or pixel, which in our works is typically chosen to be a 20-micron square. The surface brightness of the background sky and that of foreground stars must be subtracted from the raw data. The data for the huge number of pixels can be most conveniently processed by digital techniques in combination with electronic computers with large memory capacity (cf. P11).

## 2.2. Isophotal Map and Substructures

The simplest way to replace a photographic image with the surface brightness data is to represent them in the form of an isophotal map (simply called isophotes), which shows contours at selected surface-brightness levels of the object galaxy.

Figures A2.1b through A2.4b show isophotal maps of a spiral galaxy, M81, produced from the photographs in figures A2.1a through A2.4a, which were taken in different colors, *U* (ultraviolet), *B* (blue), *V* (visual) and *I* (infrared). The color discrimination is achieved by using different combinations of various photographic emulsions and filters. The relative sensitivity curves of the color bands conventionally used by the authors are shown in figure 2.1 (cf. G54).

The isophotal map in ultraviolet reveals the distribution of young hot stars within the galaxy, while the isophotal map in infrared reveals that of relatively old cool stars. Figure A2.6 through A2.14 show examples of galaxies of various morphological types. It is apparent from these figures that the spheroidal systems mainly consist of old cool stars, but spiral structures in the disk-like systems are dominated by young hot stars. Thus we realize that a galaxy is composed of stars of various ages, and that stars of different ages manifest different substructures within a galaxy. An analysis of the age sequence of substructures provides us with an information on how the galaxy was formed.

## 2.3. Isochromatic Map and Stellar Composition

The merit of observing galaxies in different colors becomes much clearer when the surface brightness in one color band is subtracted from another to yield color maps, known as isochromatic maps (or simply isochromes), such as the U–I map shown in figure A2.5. The blue region elliptical in shape in this map, which is closely associated with the spiral structures, indicates the location where formation of hot massive stars is actively going on in the present epoch, amidst an extended system of old cool stars (cf. P14).

Since the radiation of a cool star is strongly absorbed in the violet region by heavy atoms in the stellar atmosphere, one can even infer the variation of heavy-element abundance within a galaxy by precise measurement of color variation over an extended system of cool stars. Heavy atoms in the universe were produced by nuclear reactions in generation after generation of stars. Isochromatic maps are important tools for understanding the composition of the stellar population and the history of star formation in a galaxy.

To infer the masses and ages of stars composing subsystems of a galaxy is essential to relating the observed local luminosity to the mass density, which determines the dynamic structure and evolution of a galaxy. The stars burning hydrogen at their center roughly follow a mass-luminosity law $L \propto m^{\kappa}$, $k$ being about 3.5. The mass-to-luminosity ratio, $m/L$, differs by a factor of 200 between hot blue stars of 10 solar masses and cool stars similar to the sun. The ratio $m/L$, however, decreases by a factor of more than 1000 when a star of one solar mass has consumed its hydrogen at the center and evolves into a red giant. It is often difficult, however, to determine the ratio of the evolved to the unevolved cool stars from color analysis alone, and one needs the help of spectroscopic analyses.

## 2.4. Effects of Projection and Internal Absorption

An isophotal map is the projection of three-dimensional luminosity distribution onto the plane perpendicular to the line of sight. In order to recover the three-dimensional information one necessarily assumes a structural model, which will be described in detail in the following sections. A combination of a spheroidal and a disk component is the most conventional one. The observed luminosity is subject to the effect of internal absorption along the line of sight caused by a thin layer of interstellar dust lying in the equatorial plane of the disk component. The effect of this absorption, if appreciable, must be corrected in deriving the intrinsic luminosity and colors. This can be done either statistically or by assuming the distribution and the absorption characteristics of interstellar dust. Dense dust clouds often manifest themselves in the form of dark lanes, indicating regions of high concentration of interstellar matter and regions of possible star formation in a galaxy.

# 3. Radial Luminosity Distribution

## 3.1. Luminosity Profile

A graph of surface brightness versus position along a specified line drawn on the image of a galaxy is called a luminosity profile, or simply a profile. Thus a profile is a section through an isophotal map. Profiles are more widely used than isophotes in the analyses of the structure of galaxies. Most commonly used are the profiles along the major and the minor axes of the galaxy image.

What kind of information is contained in the major or minor axis profiles depends largely upon the projection effect. For galaxies seen nearly face-on, both the major-axis and the minor-axis profiles show how luminous matter is distributed in radial and azimuthal directions with respect to the rotation axis, while the distribution perpendicular to the galactic plane is integrated along the line of sight. For galaxies seen nearly edge-on, the minor-axis profile reflects the luminosity distribution perpendicular to the galactic plane, while the radial distribution is integrated. The major-axis profile of edge-on galaxies is the projection of luminosity distribution in the galactic plane although the profile is affected by strong internal absorption for most spiral galaxies.

It is often very difficult to derive relevant structural information from the observed luminosity profiles of nearly edge-on galaxies because of the strong internal absorption. Detailed photometric studies of edge-on galaxies have just been begun and will be described in section 4. A new method of analyzing the luminosity distribution of galaxies with various inclinations will be introduced in section 5.

## 3.2. Luminosity Profile of Nearly Face-on Galaxies

The prime interest in the analysis of the luminosity profile of nearly face-on galaxies lies in studying the radial luminosity distribution, that is, how the luminous matter is distributed with distance from the rotation axis of a galaxy. For this purpose, the structure perpendicular to the galactic plane is not taken into account, and galaxies are treated as if they consisted of an infinitesimally thin disk. It is also assumed that the luminosity profile is axisymmetric with respect to the rotation axis, and the small-scale bumpiness in the profile due to spiral arms etc. is regarded as secondary structure. The profile with small-scale irregularities smoothed out is more suitable for the study of the radial luminosity distribution than is the observed raw profile.

It should be kept in mind, however, that there are some distinct components in real galaxies, such as bars, lenses and rings, whose luminosity distributions are not necessarily axisymmetric (G33).

## 3.3. Smoothing of the Luminosity Profile

Two types of smoothing method are conventionally applied to the observed raw profile. One is the averaging of the profiles along various directions in the galactic plane and the other is the construction of an 'equivalent profile'.

For galaxies seen inclined, the distance from the galactic center measured in the plane of the galaxy appears shrunken because of the projection effect, except in the direction of the major axis. It is thus necessary to correct the raw data for this shrinkage when the profiles in various directions are averaged. A 'mean profile' often refers to the major-axis profile averaged over both halves with respect to the galactic center.

The equivalent profile was introduced by de Vaucouleurs (G10; G13). This was later adopted as a standard method to represent the luminosity profile by the Working Group on Galaxy Photometry and Spectrophotometry in Commission 28 of the International Astronomical Union (IAU). The equivalent profile is constructed as follows. Let $S$ be the area included in an isophote of surface-brightness level $I$. If the isophote consists of 'islands', $S$ includes the areas of all such islands. In other words, $S$ is the total area brighter than $I$. The equivalent radius, $r^*$, is defined as the radius of a circle which has the same area as $S$, that is, $r^* = (S/\pi)^{1/2}$. A plot of the surface brightness versus the equivalent radius, $I(r^*)$, is the equivalent profile.

The equivalent profile has the advantage that we can obtain a very smooth profile even for galaxies whose isophotes have considerable irregularities. The disadvantage of the equivalent profile, on the other hand, is that it is a monotonically decreasing function by definition; bumpiness with large amplitude cannot be reflected appropriately in the equivalent profile. It is also noted that the equivalent profile is affected by the contraction of the distance scale due to projection effects noted above.

Figures A3.1 through A3.6 show examples of profiles together with photographs and isophotes for galaxies of various morphological types. The dot-dashed and the broken curves are the mean profiles along the major and the minor axes, respectively, while the solid curve shows the profile averaged over the image of the galaxy in various directions, contraction of the distance scale having been corrected. The solid curve with circles shows the equivalent profile. The ordinate is the surface brightness in units of magnitude per square second of arc. The abscissa is the radial distance from the galactic center for the mean and the averaged profiles, and the equivalent radius for the equivalent profile.

## 3.4. Laws of Radial Luminosity Distribution

In terms of the structure of galaxies, elliptical galaxies and late-type spiral galaxies are two opposite extremes. Elliptical galaxies consist only of an ellipsoidal component without any flat disk component, while in late-type spiral galaxies the disk component is dominant and the central bulge component of ellipsoidal shape, if present, is very weak. Galaxies which lie between the two extremities are composite systems where the bulge and the disk coexist. The relative strength of the two components varies along the morphological sequence; generally the bulge component is strong in lenticular and early-type spiral galaxies while the disk component becomes stronger in the later morphological types. This structural difference appears as a systematic change in the luminosity profile along the morphological sequence, as demonstrated in figure 3.2. However, the observed profiles are often much more complicated than the simplified interpretation given above because of the presence of components other than the bulge and disk. Apparently, the form and color of the bulge component are very similar to those of elliptical galaxies. Accordingly, we expect to be able to understand the luminosity distributions of galaxies systematically by studying those of the two extremities, i.e., elliptical galaxies and late-type spiral galaxies (G16; C5).

The luminoisity distribution of elliptical galaxies has been investigated by many astronomers, and several analytic functions have been proposed to model the distribution. Among others the most commonly used is the '$r^{1/4}$-law' proposed by de Vaucouleurs (G10). This is written as

$$I(r) = I(0) \exp[-(ar)^{1/4}], \tag{3.1}$$

or, if the surface brightness is expressed in magnitude units we have

$$\mu(r) = \mu(0) + (a'r)^{1/4}, \tag{3.2}$$

where r is the distance from the galactic center and $I(0)$, $\mu(0)$, $a$ and $a'$ are parameters which differ galaxy by galaxy; $I(0)$ and $\mu(0)$ are the surface brightness at the galactic center ($r=0$); and $a$ and $a'$ represent the size of the galaxy with $a'=1.390a$. The $r^{1/4}$-law is expressed by a straight line in a plot of $\mu(r)$ versus $r^{1/4}$. An example is shown in the inset of figure A3.1c.

As seen in figure A3.6c, the profile of the disk component of late-type spiral galaxies closely resembles a straight line in a plot of $\mu(r)$ versus r. This indicates that the luminosity distribution of pure disk component can be modeled by the 'exponential law', written as

$$I(r) = I(0) \exp(-br), \tag{3.3}$$

or in magnitude units as

$$\mu(r) = \mu(0) + b'r, \tag{3.4}$$

where the symbols have the same meanings as for the $r^{1/4}$-law and $b'=1.086b$.

Figure 3.1 shows schematically how the two components, the $r^{1/4}$-law bulge and the exponential law disk, can produce a variety of composite profiles. A new technique to decompose an observed profile into the bulge and the disk components was introduced by Kormendy (G32). Data on the systematic properties of the components have been accumulated by applying this technique.

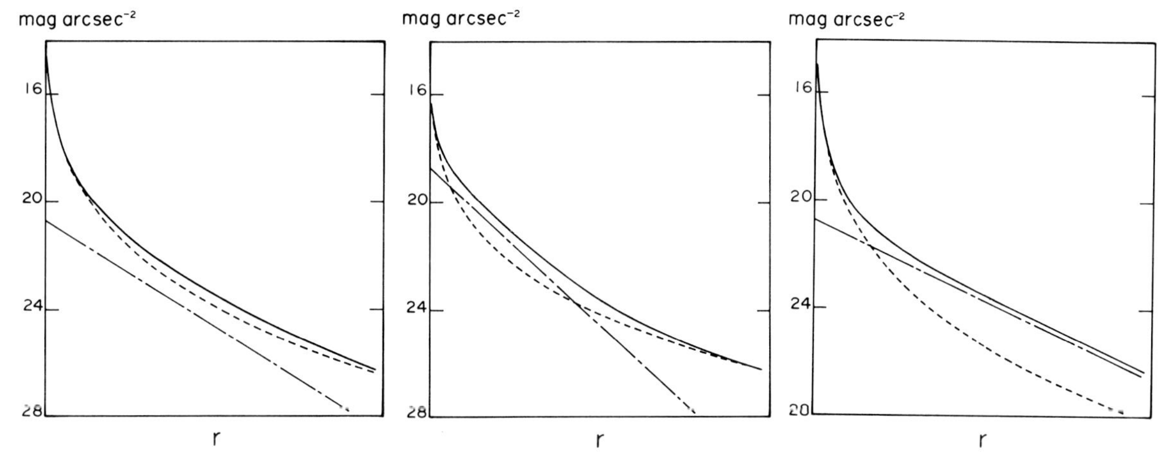


Fig. 3.1. Examples of the composite profile (solid curve) made by combining a $r^{1/4}$-law bulge (broken curve) with an exponential disk (dot-dashed line) at different relative strengths. The ordinate is the surface brightness and the abscissa is the linear distance from the galactic center in arbitrary units.

## 3.5. Normalization of the Luminosity Profile

The intrinsic sizes and surface brightnesses of galaxies cover an extremely wide range. Furthermore, the apparent size of a galaxy depends upon the distance, which is a quantity very difficult to determine. Thus the observed profiles themselves are not always suitable for the comparative studies of the forms of luminosity profiles, and normalized luminosity profiles must be obtained for that purpose.

In order to normalize a profile of a galaxy, we have to have a characteristic size and a characteristic surface brightness which can represent the galaxy. Usually the equivalent radius, $r^*_e$, and the surface brightness, $I_e$, of the isophote within which a half of the total luminosity is emitted are adopted as the characteristic parameters. The parameter $r^*_e$ is called the 'effective equivalent radius' and $I_e$ the 'effective surface brightness'.

The reduced surface brightness, $J$, and the reduced equivalent radius, $\rho^*$, are defined by

$$J = I/I_e \quad \text{and} \quad \rho^* = r^*/r^*_e. \tag{3.5}$$

A plot of log $J$ versus $\rho^*$ is called the reduced profile. This reduced profile is independent on

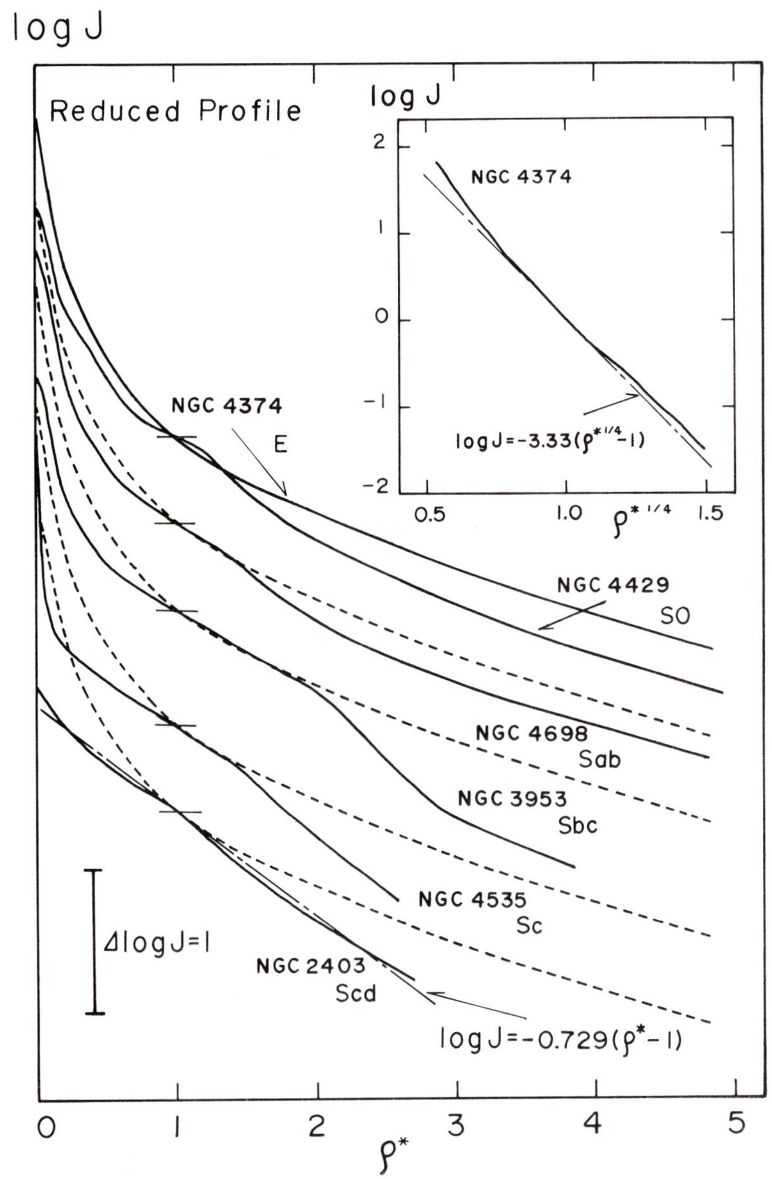


Fig. 3.2. Reduced profiles of galaxies of various morphological types. Profiles are shifted arbitrarily along the vertical ordinate. A horizontal dash indicates the level where $\log J = 0$. For convenience of intercomparison, the profile of the elliptical galaxy NGC 4374 is repeated at each profile (broken curve). The $r^{1/4}$-law (inset) and the exponential law are shown by the dot-dashed lines.

the intrinsic size, intrinsic surface brightness and apparent size. Thus we can compare the reduced profiles of different galaxies in order to investigate the shape of profiles, i.e., the luminosity distribution law.

Reduced profiles correspodning to the $r^{1/4}$-law and the exponential law are given by

$$\log J = -3.33 \, (\rho^{*1/4} - 1), \qquad (3.6)$$

and

$$\log J = -0.729 \, (\rho^{*} - 1), \qquad (3.7)$$

respectively. It should be noted that these equations are by no means the rigorous expressions for the reduced profiles of real elliptical galaxies or real disk components. These equations are useful as analytical expressions which approximate the observed profiles. Reduced profiles of the galaxies in figures A3.1 through A3.6 are shown in figure 3.2 for convenience of comparison.

# 4. Luminosity Distribution Perpendicular to the Galactic Plane

## 4.1. Surface Photometry of Edge-on Galaxies

The basic components which compose the global structure of galaxies are the halo, bulge, bar, lens, disk, and ring (G33). Galaxies have been morphologically classified according to the relative importance of these components (G12). Among these the bulge and disk are the most fundamental components (cf. section 3.4).

Investigation of the properties of these two components of galaxies has been carried out using the method of profile decomposition. In order to decompose the profiles we assume that the bulge dominates over the disk in the central part of the luminosity profile while in the outer part the disk becomes dominant. However, recently Kormendy (G32) decomposed the profiles of lenticular galaxies by means of an iterative fitting method and showed that the contributions of the bulge components to the apparent disk profiles are not negligible throughout the radius range observed. This means that the unambiguous decomposition of the profile of face-on galaxies is very difficult because of the unknown relative contribution of the two components in the total profile. On the other hand, the profile of edge-on galaxies can be decomposed less ambiguously into the bulge and disk components, since the characteristics of these components are most distinctive along the z-axis (the rotational axis).

An initial trial to decompose the luminosity profile of edge-on galaxies was made by Tsikoudi (G56; G57). However, she only decomposed the major- and minor-axis profiles independently. In order to perform the profile decomposition of edge-on galaxies properly, the application of a three-dimensional model of the volume emissivity distribution is essential. We have developed fundamental techniques to analyze the global three-dimensional structure of galaxies, as shown in the section 4.3 (P2-P5). Similar analyses of edge-on spiral galaxies were carried out by van der Kruit and Searle (G62-G65), and the structure of disk components was investigated.

## 4.2. Features Seen in Edge-on Galaxies

Some peculiar features such as a box-shaped bulge, warping of the disk and a thin layer of strong absorption are observed best in edge-on galaxies. Figure A4.8 shows a lenticular galaxy, NGC 7332, which has a box-shaped bulge (G15). Warping of the disk can be seen in NGC 4762 (figures A4.2a and 2b; P4) and NGC 4565 (figures A4.5a and 5b; P3). Warping of the disk in these galaxies is seen also in the distribution of neutral hydrogen. NGC 891 is a typical example of a galaxy with a strong dark lane. Complex features of the dark lane are clearly seen in figure A4.16. These internal absorptions are often serious problems in studies of edge-on galaxies.

It should be noted that the morphological classification of edge-on galaxies is highly ambiguous because we cannot see details of structures in the galactic plane such as spiral arms. The main criteria for the morphological classification of edge-on galaxies are the presence of a dark lane in case of lenticular galaxies and the bulge-to-disk ratio in case of spiral galaxies. Figures A4.1 through A4.18 show edge-on galaxies of various types.

## 4.3. Luminosity Distribution of Bulge and Disk Components

NGC 5866 is a lenticular galaxy with a dominant bulge and is an object suitable for the investigation of bulge structure. The photograph, isophotes and luminosity profiles along the major and the minor axes are shown in figures A4.1a through 4.1d. The minor-axis profile is well approximated by de Vaucouleurs' $r^{1/4}$-law, which indicates its similarity to elliptical galaxies (P2).

In contrast, NGC 4762, a lenticular galaxy with prominent disk is an ideal object for studying the disk component because of its exactly edge-on aspect and the lack of an absorption layer. Figures A4.2a through A4.2c show the photograph, isophotes and the luminosity profiles perpendicular to the galactic plane of this galaxy. As shown in figure A4.2c, the luminosity of the disk component decreases with z more slowly than the exponential law which simple isothermal models predict. Furthermore, the major axis profile of this galaxy shows two well-defined humps at $r \sim 30''$ and $90''$. They suggest the existence of other two basic components (P6; P18).

## 4.4. Three-Dimensional Models of Galaxies

We have introduced simple three-dimensional semi-empirical models in order to decompose the luminosity profiles of edge-on galaxies into the bulge and the disk components (P5). Assuming that galaxies are axisymmetric, we adopt the following models of the volume emissivity distribution:

$$\rho_S (x, y, z) = a^{\gamma+1} \rho_{0S} (a^2 + x^2 + y^2 + z^2/c^2)^{-(\gamma+1)/2}, \qquad (4.1)$$

for the bulge component and

$$\rho_D (x, y, z) = \rho_{0D} \exp[-\alpha (x^2 + y^2)^{1/2} - \beta |z|], \qquad (4.2)$$

for the disk component, where (x, y, z) is a Cartesian coordinate system in which z is taken parallel to the rotational axis, and $a$ (measure of the bulge size), $c$ (axial ratio of bulge), $\gamma$ (luminosity gradient of bulge), $\alpha^{-1}$ (radial scale-length of disk), and $\beta^{-1}$ (scale-thickness of disk) are parameters. On the basis of these models we compute luminosity profiles along several directions for edge-on view assuming no internal absorption. The resulting profiles are fitted to the corresponding observed profiles of a galaxy to yield appropriate model parameters.

Care is taken in the model fitting procedures to avoid the effects of internal absorption. By this model fitting we have derived the model parameters for several edge-on galaxies of various morphological types (P5). Using the model parameters we can produce a hypothetical profile of each component and of the galaxy which would be observed if the galaxy were to be seen face-on. Figures A4.3 through A4.6 show photographs, isophotes and the rectified face-on profiles of typical edge-on galaxies, NGC 3115 (S0), NGC 4594 (Sa), NGC 4565 (Sb) and NGC 4244 (Scd). The dominance of the bulge relative to the disk in the face-on profile varies systematically along the sequence of morphological types (P4; P5).

The parameter $\gamma$ in equation (4.1) is a measure of the luminosity concentration in the bulge component. Figure 4.1 shows $\gamma$ plotted against the morphological type. The luminosity concentration becomes stronger along the morphological sequence from ellipticals through lenticulars to spirals. This continuous change in $\gamma$ supports the hypothesis that lenticulars are galaxies transitional in type between elliptical and spiral galaxies (P5).

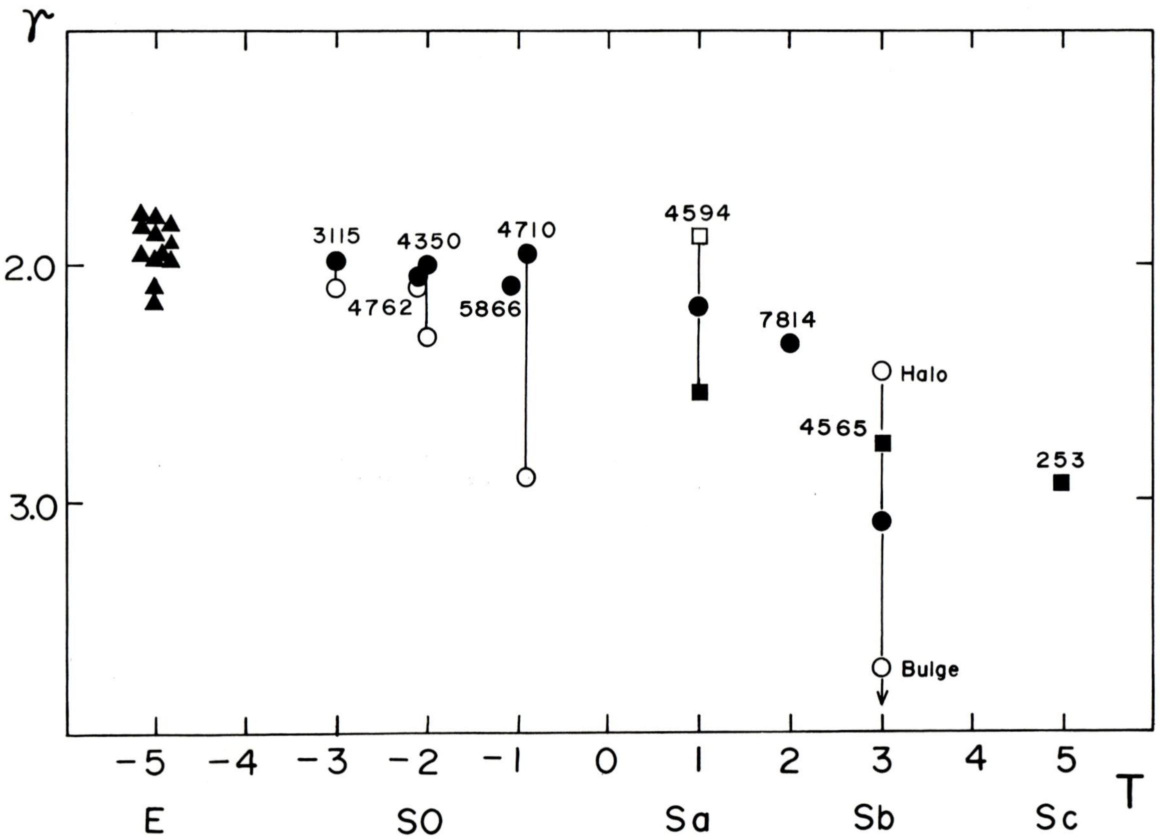


Fig. 4.1. Luminosity gradient $\gamma$ plotted against the morphological type index $T$. Different symbols correspond to different observers or different methods of analysis.

# 5. Projected Profile and Related Analysis

## 5.1. Projected Profile and Generalized Radial Profile

The projected profile $f(x)$ was introduced for the purpose of comparative studies of the global structure of galaxies at various inclination (P19; P20). The projected profile is calculated by projecting the surface brightness distribution $s(x, y)$ onto the major axis of the image of a galaxy. The function $f(x)$ is defined as,

$$f(x) = \int_{-\infty}^{\infty} s(x, y)\, dy, \tag{5.1}$$

where $x$ is measured along the major axis. The integrand $s(x, y)$ is the surface brightness. This calculation is performed by the computer using digital techniques.

If galaxies are assumed to be optically thin, $f(x)$ is independent of the inclination of the galaxy. Moreover, if galaxies are assumed to be axisymmetric, $f(x)$ can be transformed into the generalized radial profile $\sigma(r)$. The function $\sigma(r)$ is related to $f(x)$ through the integral equation

$$f(x) = 2 \int_{|x|}^{\infty} (r^2 - x^2)^{-1/2}\, \sigma(r) r\, dr, \tag{5.2}$$

where $r$ is the distance from the galactic center when the galaxy is seen face-on. The integral equation is solved numerically to yield $\sigma(r)$. The generalized radial profile is the profile which would be observed if the galaxy were to be optically thin, rotationally symmetric and seen face-on.

The profiles $f(x)$ and $\sigma(r)$ have a potential advantage over major-axis profiles or equivalent profiles for comparing the characteristics of galaxies having different inclinations. It must be kept in mind, however, that the internal absorption and non-axisymmetry of real galaxies affect the profiles $f(x)$ and $\sigma(r)$ producing spurious features. These profiles may be used most successfully in the infrared surface photometry of galaxies.

## 5.2. Representative Examples

Six representative examples are shown in figures A5.1 through A5.6, where direct photographs, isophotes, the projected profiles $f(x)$ and the generalized radial profiles $\sigma(r)$ are presented. NGC 4374, an elliptical galaxy, has very smooth profiles which gently curve upward. NGC 4267, a face-on lenticular galaxy, resembles an elliptical galaxy in appearance, but its $f(x)$ profile shows a sudden change in gradient, indicating the presence of both bulge and disk components. NGC 4026 is an edge-on lenticular galaxy. Its profiles resemble those of NGC 4267 in spite of different photographic appearance. NGC 4569 is an early-type spiral galaxy with spiral structures of low surface brightness. Its $f(x)$ profile curves gently downward, and the $\sigma(r)$ profile shows definite signs of two-component structure, that is, a bulge and an exponential disk. NGC 4100, a late-type spiral galaxy, has bright well-developed spiral structures. The $f(x)$ profile curves downward strongly, and the $\sigma(r)$ profile has a very bumpy appearance caused by the spiral structure. NGC 4234 is one of the latest-type galaxies. The profiles suggest that its bulge component is practically non-existent, while its disk region resembles that of NGC 4100.

## 5.3. Quantitative Expression of the Global Radial Structure

These six examples well represent the systematic variation of the profiles with morphological type. In general the transition point (i.e., the point where the gradient of the $f(x)$ profile changes suddenly as noted above) first appears in the $f(x)$ of the S0 class, and then shifts progressively inward in later types. The curvature of the $f(x)$ profile in the major outer region of a galaxy varies from upward for elliptical galaxies to downward for late-type spirals. This reflects the well-known systematical change in the structure of galaxies described in section 3.4. The various characteristics of the projected profiles can be represented by parameters derived through appropriate mathematical treatments to $f(x)$ or $\sigma(r)$. One of them is the introduction of the 'mean concentration index'. A series of central concentration indices is defined as

$$P_n - 100 \int_0^{r_{26}/2^n} r\sigma(r)\, dr \Big/ \int_0^{r_{26}} r\sigma(r)\, dr, \qquad (5.3)$$

where $r_{26}$ is the radius where $\sigma(r)$ corresponds to 26 mag arcsec$^{-2}$. Five indices $P_n$ ($n$=1–5) are computed for some 250 galaxies and found to correlate with each other very well. We therefore define a single mean concentration index, $X1(P)$, by the mean value of $P_n$($n$=1–5) in normalized form, i.e., with a mean of zero and a dispersion of unity (P15). The correlation between $X1(P)$ and the morphological type index, $T$, given in the RC2 catalogue (C5) is shown in figure 5.1 for 200 galaxies (P15). The good correlation suggests that the central concentration might be a primary factor that determines the morphological type.

The parameter $X1(P)$ is given for selected galaxies, together with photographs, in figures A5.7 through A5.36, arranged in order of decreasing $X1(P)$ index. Other examples are presented in a recent publication (P15; P20).

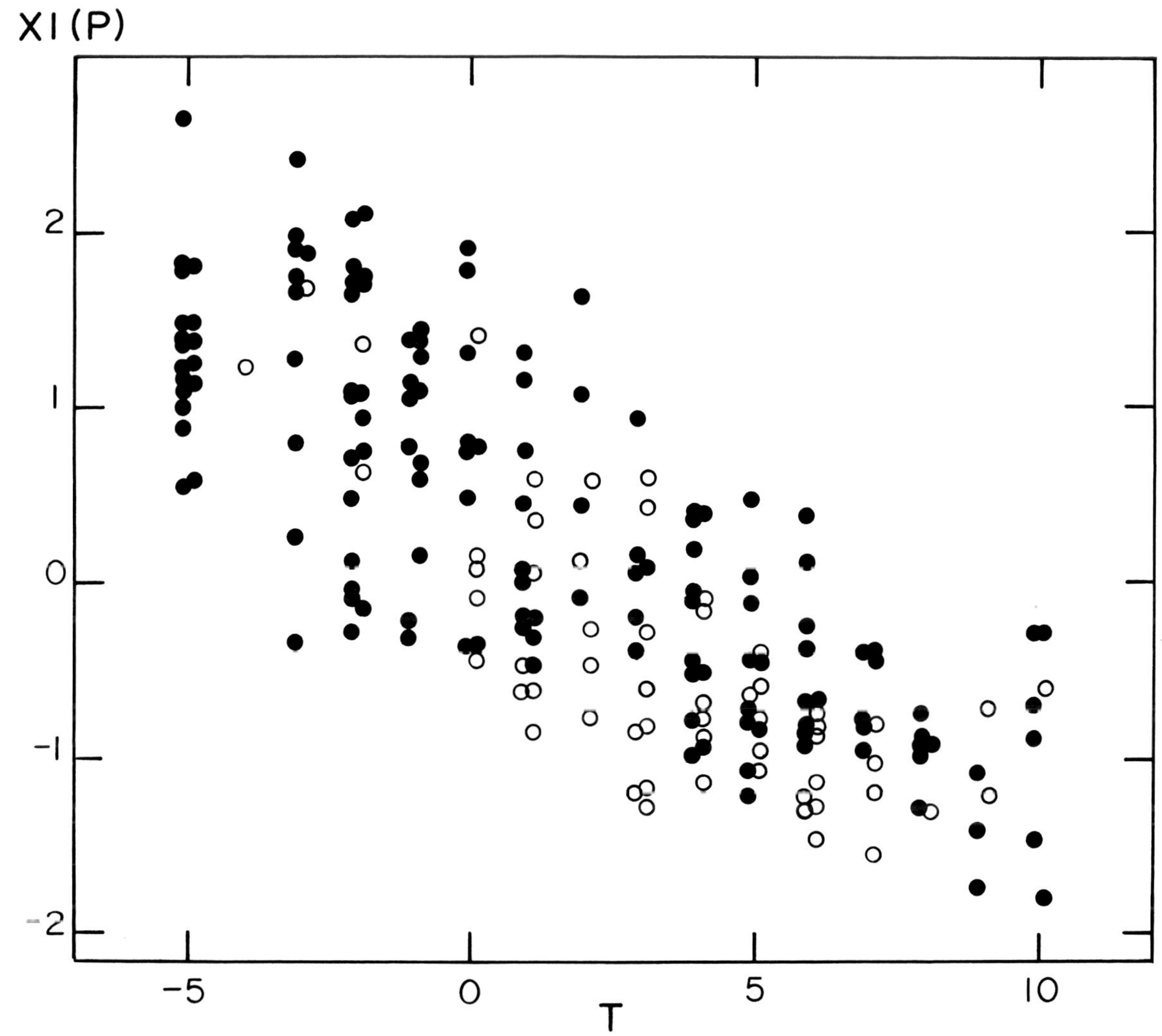


Fig. 5.1. The correlation between the mean concentration index, $X1(P)$, and the morphological type index, $T$, given in RC 2 catalogue for 200 galaxies. The correlation coefficient is -0.79.

# 6. Spectral Analysis of Spiral Patterns

## 6.1. The Background and Objectives of the Analysis

Spiral structures are visually the most impressive features in the morphology of galaxies. Most of these spiral structures can reasonably be well understood as the manifestation of global modes of oscillation of galactic disks (G6; P1). The identification and discrimination of these modes is important for the study of dynamical properties of observed galaxies. The method of two-dimensional Fourier analysis presented in this section gives a quantitative basis to achieve this objective.

Preliminary application of the present method to M51 is reported by Iye et al. (P7), and a full description of the technique and results of application to NGC 4254 are given by Iye et al. (P8). They found that the asymmetric spiral structure of NGC 4254 is delineated as a superposition of five-armed and three-armed spiral components on the main one-armed spiral component. They suggested that the origin of the asymmetric spiral structure of NGC 4254 must be explained without the help of spiral exciters such as companion galaxies or a bar. This is because they could not find such exciters in the case of NGC 4254, and even if an irrecognizably weak exciter were present, it could not produce such an asymmetric spiral structure. Other works involving Fourier analysis have been performed on the spiral pattern of M31 by Kalnajs (G29) and by Simien et al. (G52). They found a one-armed leading spiral component as the principal component of M31 instead of the traditionally assumed two-armed trailing spiral component. These findings clearly show the importance and usefulness of the Fourier analysis for objective pattern recognition.

## 6.2. Fourier Analysis of the Observed Intensity Distribution

First the relevant values of the position angle of the major axis and the inclination of a galaxy are determined by some objective method, and the observed relative intensity distribution $I_R(x_0, y_0)$ is transformed to a face-on intensity distribution $I_F(x, y)$. The second step is to choose an annular region in the plane of the galaxy ($r_{min} < r < r_{max}$) to be analyzed, where $r = (x^2+y^2)^{1/2}$. In this atlas, $r_{max}$ and $r_{min}$ are defined as the radii where the azimuthal mean surface brightness of the galaxy becomes 25 mag arcsec$^{-2}$ and 20 mag arcsec$^{-2}$ respectively. Then the face-on intensity distribution $I_F(x, y)$ is transformed into an intensity distribution $I(u, \theta)$ in logarithmic polar coordinates, where $u = \ln r$ and $\theta = \tan^{-1}(y/x)$. $I(u, \theta)$ is separated into an axisymmetric part $I_0(u)$ and a non-axisymmetric part $I_1(u, \theta)$. The axi-symmetric part $I_0(u)$ is regarded as the intensity distribution of the basic disk while $I_1(u, \theta)$ is thought of as a perturbation superimposed on $I_0(u)$ to give spiral structure or whatever.

The intensity distribution $I(u, \theta)$ of a galaxy can be expressed as a superposition of many logarithmic spiral components, that is,

$$I(u, \theta) = \sum_{m=0}^{\infty} \int_{-\infty}^{\infty} A(\alpha, m) \exp[i(\alpha u + m\theta)]\, d\alpha, \tag{6.1}$$

where $m$ is the azimuthal wave number or the number of spiral arms and $\alpha$ is the radial wave number or a measure of the tightness of spiral winding, its sign showing the winding direction. The Fourier coefficient $A(\alpha, m)$ represents the significance of the spiral component with wave numbers $\alpha$ and $m$. The decomposition of the observed pattern into a number of principal spiral components can be accomplished by evaluating the power spectrum of the amplitude $|A(\alpha, m)|$. The normalized power spectrum of the non-axisymmetric component $I_1(u, \theta)$ is defined by the expression

$$P(\alpha, m) = |A(\alpha, m)|^2 / \sum_{m=1}^{\infty} \int_{-\infty}^{\infty} |A(\alpha, m)|^2\, d\alpha. \tag{6.2}$$

Figures A6.1a through A6.1d illustrate the procedure and results of the analysis of NGC 4535. Figure A6.1a is a photograph which gives an idea of the observed relative intensity distribution $I_R(x_0, y_0)$, and figure A6.1b shows $I(u, \theta)$. Figure A6.1d presents the normalized power spectrum $P(\alpha, m)$, where the phase angle $\phi = \arg[A(\alpha, m)]$ is also shown. Figure A6.1c shows the face-on image reconstructed from all of the logarithmic spiral components with $P(\alpha, m) \geqq 0.01$, which are indicated by the hatched area in figure A6.1d.

## 6.3. Indices Related to Morphological Features

Some of the morphological features which have been defined so far by subjective pattern recognition could be quantitatively defined by the proper use of the power spectrum. For example, the significance of the bar-like component can be evaluated by an index,

$$B_{bar} = \int_{-\pi/2}^{\pi/2} P(\alpha, 2)\, d\alpha. \tag{6.3}$$

Table 6.1. Morphological indices and the parameters adopted in the analysis.

| Galaxy | Type | $S_{ex}(2)$ $\times 10^2$ | $B_{bar}$ $\times 10^2$ | $R_{odd}$ $\times 10^2$ | $\theta$ (deg) | $i$ (deg) | $r_{max}$ (arcsec) | $r_{min}$ |
|---|---|---|---|---|---|---|---|---|
| NGC 4535 | SAB(s)c | 31.9 | 39.5 | 12.9 | 16 | 47 | 200 | 14 |
| NGC 4501 | SA(rs)b | 9.4 | 8.5 | 64.1 | 138 | 62 | 177 | 20 |
| NGC 4579 | SAB(rs)b | 3.7 | 19.2 | 2.0 | 103 | 38 | 225 | 26 |
| NGC 4321 | SB(rs)bc | 3.2 | 6.0 | 71.3 | 72 | 31 | 264 | 15 |
| NGC 4340 | SB(r)0$^+$ | 1.5 | 28.9 | 3.8 | 91 | 45 | 94 | 10 |
| NGC 4472 | E2 | -0.4 | 32.0 | 1.9 | 152 | 38 | 392 | 35 |
| NGC 4303 | SAB(rs)bc | -16.5 | 21.0 | 15.9 | 107 | 32 | 192 | 13 |

The global tendency of spiral winding direction of two-armed spiral components and their degree of regularity can be measured by a parameter,

$$S_{ex}(2) = \int_0^{\infty} P(\alpha, 2)\, d\alpha - \int_{-\infty}^{0} P(\alpha, 2)\, d\alpha. \tag{6.4}$$

The degree of asymmetry of spiral structure can be represented by an index,

$$R_{odd} = \sum_{m=1}^{\infty} \int_{-\infty}^{\infty} P(\alpha, 2m-1)\, d\alpha \Big/ \sum_{m=1}^{\infty} \int_{-\infty}^{\infty} P(\alpha, m)\, d\alpha, \tag{6.5}$$

and so on. Table 6.1 summarizes the morphological feature indices defined above and the parameters adopted in the analysis for seven selected galaxies which are presented in figures A6.1 through A6.7 in order of decreasing $S_{ex}(2)$.



# 7. Active Galaxies and Peculiar Galaxies

## 7.1. Seyfert Galaxies, N Galaxies and Quasars

In 1943 Seyfert (G49) studied several galaxies characterized by bright stellar nuclei which show spectra with strong broad emission lines. Later the number of objects of this kind increased, and they came to be called Seyfert galaxies. In 1958 Morgan (G38) isolated the galaxies having small brilliant nuclei and named them N galaxies. A highlight of the early 1960s was the discovery of quasars and the identification of their highly redshifted spectral lines which suggested that these might be extremely luminous extragalactic objects. Since the spectral features of some types of Seyferts and quasars are quite similar, Seyferts' nuclei are regarded as mini-quasars. Seyferts, quasars, and N galaxies all seem to possess active nuclei which are in various states of excitement. A quasar, Q1216+158, and its spectrum are shown in figures A7.1a and 1b (P10), while several Seyferts are in A7.2 through A7.4.

## 7.2. Survey for Ultraviolet-Excess Galaxies and Emission-Line Galaxies with Schmidt Telescopes

Systematic searches for ultraviolet-excess galaxies and emission-line galaxies have been carried out on objective prism plates taken with Schmidt telescopes. In 1967 Markarian began to publish results of the observation which he had made using the Byurakan 1-m Schmidt with a 1.5° prism. Up to 1981 Markarian et al. (G35) listed 1500 galaxies which have a strong ultraviolet continuum over a sky area of 15,000 square degrees. They are called Markarian galaxies, and about 10% of them are known to have characteristics of Seyfert galaxies or quasars. Kazaryan and Kazaryan (G30) is continuing the same kind of survey and its publication at the Byurakan Observatory.

With the 105-cm Kiso Schmidt telescope of the Tokyo Astronomical Observatory we have been conducting a survey for ultraviolet-excess galaxies, termed KUG (Kiso UV-excess Galaxy), by means of the multi-exposure method in three color bands of $U$, $G$ and $R$ (cf. P16). This is a modified application of the $UBV$ three-color photograph attempted by Haro (G19) in 1956, using the 66-cm Tonantzintla Schmidt telescope. At that time he found 44 UV-bright galaxies, which were called Haro galaxies.

Markarian galaxies and KUGs are essentially the same since both groups are galaxies selected by the criterion that their luminosity in the blue-violet region of the wavelength is brighter than that in the yellow-red region. In fact, follow-up observations of several KUGs have revealed that many of them have emission-line spectra and various morphological peculiarities (cf. P17). Several KUG specimens are shown in figures A7.5 through A7.9. The bright quasar survey was carried out as a subset of the Palomar-Green survey of ultraviolet-excess stellar objects by means of two-color photography in $U$ and $B$ bands with the 46-cm Schmidt telescope of the Palomar Observatory. Through the slit spectroscopy for blue objects thus detected, nearly one hundred quasars to an average limiting magnitude of $B \sim 16$ mag were later found over an area of 10,700 square degrees by Schmidt and Green (G51).

Surveys for emission-line galaxies have been conducted using the 61-cm Curtis Schmidt at the Cerro Tololo Observatory and the 60-cm Burrell Schmidt now at the Kitt Peak Observatory, both with objective prisms (cf. G31; G44). Quasar candidates detected by the Tololo survey were subsequently observed with a slit spectrograph, and some of them were confirmed as quasars (G43). A University of Michigan group carried out a Michigan-Tololo survey and found many emission-line galaxies, quite a portion of which were quasar candidates (G37). Quasar surveys are now in progress by the United Kingdom 1.2-m Schmidt team to a faint limiting magnitude of 20.5 mag using a thin objective prism with low dispersion (G47).

## 7.3. Peculiar Galaxies

Many peculiar galaxies including interacting galaxies have been found on the Palomar Sky Survey prints by Vorontsov-Velyaminov (G66; G67) and compiled in his 'Atlas of Interacting Galaxies'. Parts I and II of this atlas, containing 355 and 1500 objects, respectively, were published in 1959 and 1977. In 1966 Arp (C1) issued an atlas of high magnification photographs of 338 peculiar galaxies taken with the 5-m Hale telescope. Compact galaxies, many of which are ultraviolet-bright, and post-eruptive galaxies which show peculiar features such as jets, flares, and so on, are collected in Zwicky's Catalogue published in 1971 (C11).

On the basis of several morphological studies of Markarian galaxies and KUGs it is very likely that a fairly large portion of the UV-excess galaxies are more or less peculiar in their appearance (P17). Figures A7.10 and 11 show examples of peculiar and interacting galaxies, respectively.

# 8. Clusters of Galaxies

## 8.1. Richness and Morphological Classifications

Various concentrations of galaxies are observed, from poor groups to rich clusters. The study of clusters of galaxies is important both for understanding their intrinsic natures as fundamental systems where galaxies are formed and evolve, and for investigating the large-scale structure of the universe. Richness class, a measure of the number of member galaxies in a rich cluster, was defined by Abell (G1) according to the number of galaxies within a fixed magnitude range and a fixed radius from the cluster center as follows:

| Richness Class | Number of Galaxies |
|---|---|
| 0 | 30 - 49 |
| 1 | 50 - 79 |
| 2 | 80 - 129 |
| 3 | 130 - 199 |
| 4 | 200 - 299 |
| 5 | 300 - |

Poor clusters have substantially fewer than thirty members.

Morphological classifications for rich clusters were proposed by Bautz and Morgan (G7; hereafter referred to as BM), Rood and Sastry (G45; hereafter RS), and Struble and Rood (G53; hereafter revised RS). The BM classification is based upon the luminosity contrast of the brightest galaxy relative to other member galaxies in a cluster, having the three main classes as follows:

Type I: Clusters containing a centrally located cD galaxy.
Type II: Clusters whose brightest galaxy or galaxies are intermediate in appearance between class cD and Virgo-type giant ellipticals.
Type III: Clusters containing no dominant galaxies.

The RS and revised RS classifications are based upon the apparent spatial distribution of the ten brightest galaxies within a cluster. The main criteria of the revised RS classification are as follows:

Type cD (supergiant, dominant galaxy)
There is a dominant, generally centrally located galaxy with an extended and elongated envelope; defined in practice by having a semi-major axis which is more than 3 times that of any other cluster galaxy.
Type B (supergiant binary)
There are two supergiant galaxies separated by less than 10 times the diameters of

the larger galaxy and with a size which is more than 3 times that of any cluster member.

Type C (core-halo)
Four or more of ten brightest galaxies have comparable separations and are centrally located among numerous fainter galaxies.

Type L (line)
Four or more of ten brightest galaxies are arranged approximately colinearly with numerous fainter members located around them.

Type I (irregular)
Galaxies are distributed irregularly, often without a well-defined center.

Type F (flat)
Ten brightest galaxies and a large fraction of fainter galaxies are arranged in a flattened configuration.

Other classifications, such as those of Zwicky (C12-C17) and Oemler (G42), and correlations among different classifications are summarized in table 8.1.

Table 8.1 Summary of classification systems for clusters of galaxies.

| Symmetry<br>Central Concentration | Spherical<br>High | (Intermediate)<br>Moderate | Irregular<br>Low |
|---|---|---|---|
| Zwicky type | Compact | Medium-compact | Open |
| BM type | I, I-II, II | (II), II-III | (II-III), III |
| RS type | cD, B, (L, C) | (L), (F), (C) | (F), I |
| Oemler type | Elliptical-rich | Spiral-poor | Spiral-rich |

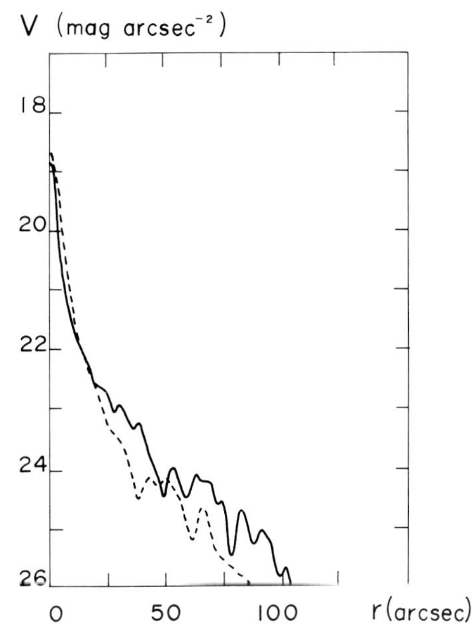


Fig. 8.1.
The luminosity profile of a suspected cD galaxy in a poor cluster MKW 1 (broken curve) and that of the brightest cD galaxy in the Coma Cluster (solid curve).

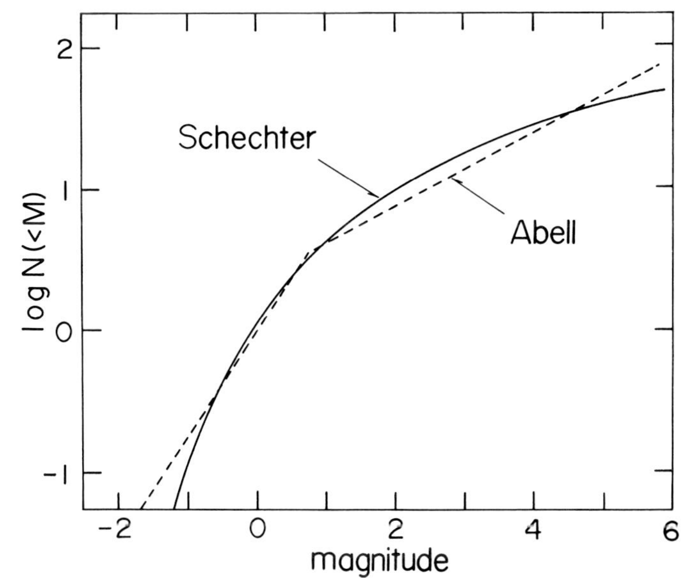


Fig. 8.2.
A comparison of the two analytic expressions of the luminosity function proposed by Abell and Schechter.

## 8.2. Clusters Including cD Galaxies

The cD galaxies which play an important role in the above classifications are defined as extraordinarily large D galaxies. A D galaxy was defined by Mathews et al. (G36) as a galaxy having an elliptical galaxy-like nucleus surrounded by an extended envelope. According to Hausman and Ostriker (G20), cD galaxies are supposed to grow through mergers of adjacent galaxies; thus, any cluster of galaxies might still undergo further evolution. Many well-defined cD galaxies are found to be located in rich clusters of galaxies. However, recently the possible presence of cD galaxies in some poor clusters is also suspected by Morgan, Kayser and White (G41: MKW) and by Albert, White and Morgan (G3: AWM). Figure 8.1 shows the luminosity profile of suspected cD galaxy in the poor cluster MKW 1, and that of the brightest galaxy of the Coma (rich) cluster. Comparison of these two galaxies shows that the former is less extended than the latter. This suggests a correlation between the size of the central galaxy and the richness of a cluster. Several clusters of galaxies representing various types are shown in figures A8.1 through A8.7.

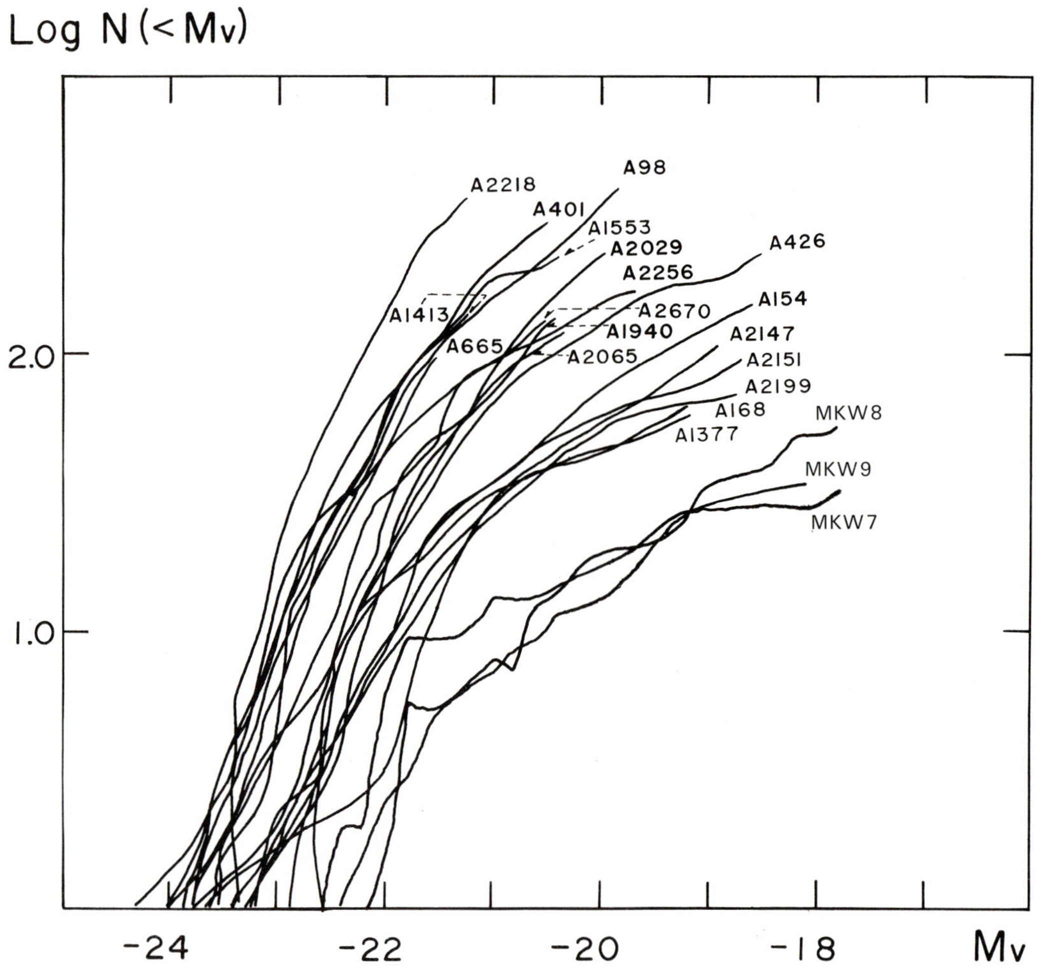


Fig. 8.3. A comparison of the luminosity functions between rich Abell clusters and poor MKW clusters.

## 8.3. Luminosity Function of Cluster Member Galaxies

The luminosity function, which is the number of member galaxies as a function of magnitude or luminosity, is a useful tool for investigation of the photometric properties of clusters of galaxies.

Abell (G2) proposed an analytical expression for the integral luminosity function of the form

$$\log N(<m) = \begin{cases} 0.75m + b_1 & (m \leqq m^*) \\ 0.25m + b_2 & (m > m^*) \end{cases} \qquad (8.1)$$

where $N(<m)$ is the number of galaxies which are brighter than or equal in brightness to the apparent magnitude, $m$, $b_1$ and $b_2$ are parameters, and $m^*$ is called the characteristic magnitude corresponding to the turning point which can be seen in the observed luminosity functions of most clusters. Later Schechter (G48) found that the analytical function,

$$N(L)\,dL = N^*(L/L^*)^{\alpha} \exp(-L/L^*)\, d(L/L^*), \qquad (8.2)$$

with $\alpha \sim -1.25$ approximates the observed differential luminosity function very well, where $L$ is the luminosity and $N^*$ and $L^*$ are parameters. The parameter $N^*$ is a measure of the richness, and $L^*$ is the characteristic luminosity which corresponds to $m^*$ in equation (8.1).

These two analytic formulae for the luminosity function are identical in practice except at the brightest end (figure 8.2). Turner and Gott (G58) found that the luminosity functions of groups of small numbers of galaxies fit well to the Schechter function with a value of $\alpha \sim -1.0$.

The integrated luminosity functions of MKW poor clusters are compared with those of Abell's rich clusters in figure 8.3. The luminosity function seems to form a continuous sequence from rich Abell clusters through poor Abell clusters to MKW clusters (cf. P22). This finding is consistent with Bahcall's argument (G5) that there exists a smooth continuation from rich to poor clusters in terms of the number of member galaxies, density and fraction of spiral galaxies.

# Appendix. Toward a Quantitative Classification

The current status of our work toward a quantitative classification is summarized briefly in this appendix. A large homogeneous sample of photometric data is required for the study of quantitative classification. We recently obtained the two-dimensional luminosity distribution in the V band for some 250 galaxies in the Virgo Cluster and the Ursa Major Clouds on the basis of plates taken with the Kiso 105-cm Schmidt telescope. The close coincidence of the distance between the two groupings greatly contributes to reducing the uncertainty in distance inherent to extragalactic works dealing with a large sample. This fact makes the present sample by far the largest set of highly homogeneous data in detailed surface photometry of galaxies available in the world at present.

Various photometric parameters were derived for the galaxies in the sample using the generalized radial profile (cf. section 5.1) within a limiting surface brightness of 26 mag arcsec$^{-2}$. They included the diameter, $D_{26}$, integrated magnitude, $V_{26}$, mean surface brightness, *SB*, and mean concentration index $X1(P)$ (see section 5.3).

As suggested by the good correlation with morphological type shown in figure 5.1, the mean concentration index is found to be sensitive to the profile shape (P15). However, since $X1(P)$ is defined on the basis of the isophotal radius, $r_{26}$, galaxies with the same profile shape but different surface brightness have different $X1(P)$. It is thus necessary to use *SB* together with $X1(P)$ to estimate the profile shape of a galaxy more accurately. Figure A.1 shows $X1(P)$ versus *SB* diagram for 143 selected galaxies of good data quality in our sample. The solid and broken curves with ticks represent model galaxy sequences following the $r^{1/4}$-law and exponential law, respectively. The curves of the two model sequences are seperated clearly in the diagram and appear to lie near the upper and lower boundaries of the distribution of the 143 galaxies. Elliptical galaxies and galaxies of latest types roughly follow the $r^{1/4}$-law and exponential sequences, respectively. This suggests that the basic

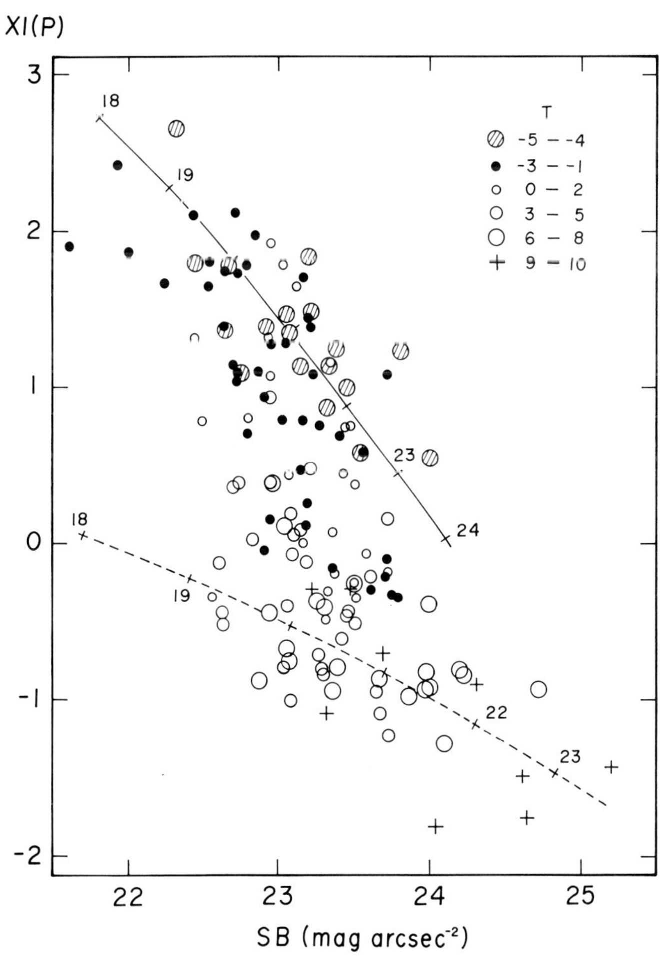


Fig. A.1.
The diagram of the mean concentration index, $X1(P)$, versus the mean surface brightness, *SB*, for 143 galaxies. The solid curve and broken curves with ticks show the $r^{1/4}$-law and exponential sequences, respectively. The ticks on the curves indicate an interval of 1 mag in the brightness-scale parameter, numerals on the ticks indicating the parameter value, $I_e$ in equation (3.5) for the solid curve and $I(0)$ in equation (3.3) for the broken curve, in units of mag arcsec$^{-2}$.

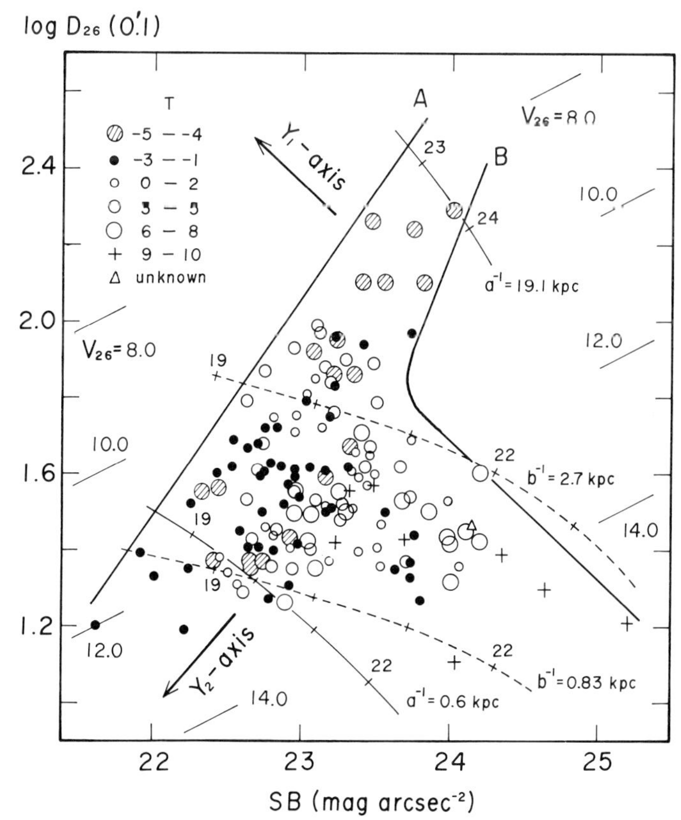


Fig. A.2.
The diameter versus surface brightness diagram for 149 galaxies in the Virgo Cluster and the Ursa Major Clouds. Projections of the axes of the two significant principal components and the lines of constant magnitude are shown. The thick solid curves (A and B) show the boundaries of the observed distribution. The thin solid and broken curves represent some model sequences for the $r^{1/4}$-law and exponential law, respectively. The meanings of the ticks and the numerals on the curves are the same as in Fig. A.1.

luminosity structure of a galaxy, especially the bulge-to-disk ratio, the most fundamental quantity of galaxy structure, can be estimated with reasonable accuracy by its position in the $X1(P)$ versus $SB$ diagram. It is noted that both $X1(P)$ and $SB$ are distance-independent. This $X1(P)$ versus $SB$ diagram may be used as a tool to estimate the luminosity structure of a distant galaxy. One important application of this method would be the characterization of clusters of galaxies according to the galaxy content.

In order to study the basic physics behind the structure and formation mechanism of galaxies, it is essential to know how many independent parameters are present. The principal component analysis of the parameter set, [log $D_{26}$, $V_{26}$, $SB$, $X1(P)$], for our sample galaxies shows that the number of the significant independent variables, *i.e.*, principal components, is two (P9; P21). They are denoted as $Y_1$ and $Y_2$. Then, the distribution of galaxies in the ($Y_1$, $Y_2$) plane must contain important information on the structure and formation mechanism of galaxies. We find that projections of the two original parameter axes, log $D_{26}$ and $SB$, onto the ($Y_1$, $Y_2$) plane are nearly orthogonal. Accordingly, we prefer, for simplicity of interpretation, the two prameters log $D_{26}$ and $SB$ as relevant quantities to characterize a galaxy. Figure A.2 shows the log $D_{26}$ versus $SB$ diagram, which we call the diameter versus surface brightness diagram (DSBD), for 149 selected galaxies in our sample (P9). The curves of model sequences corresponding to those in figure A.1 are shown for various length-scale parameters, *i.e.*, $a^{-1}$ or $b^{-1}$ (see section 3.4). We note the following in the DSBD: (1) There are two well defined boundaries (lines A and B) which limit the distribution of galaxies in addition to the spurious boundary at log $D_{26}\sim1.3$ due to the limit of observation. (2) Within the boundaries the distribution forms two distinct branches that run almost parallel to the projected $Y_1$ and $Y_2$ axes, respectively. (3) The branch along the projected $Y_2$ axis is formed mainly by ellipticals and lenticulars while that along the projected $Y_1$ axis consists of spirals, especially late-type spirals and irregulars.

The various characteristics found in the DSBD and the $X1(P)$ versus $SB$ diagram suggest the presence of some constraints on physical conditions in the present galaxies and/or in the matter from which galaxies were formed. We propose to use the DSBD and the $X1(P)$ versus $SB$ diagram as the basic tools of quantitative classification to investigate the nature of galaxies and clusters of galaxies.

# References

## General References


G 1: Abell, G. O. 1958, *Astrophys. J. Suppl.*, **3**, 211.
G 2: Abell, G. O. 1975, in *Galaxies and the Universe, Stars and Stellar Systems*, Vol. 9, ed. A. Sandage, M. Sandage, and J. Kristian (The University of Chicago Press, Chicago), p. 601
G 3: Albert, C. E., White, R. A., and Morgan, W. W. 1977, *Astrophys. J.*, **211**, 309.
G 4: Bahcall, N. A. 1977, *Ann. Rev. Astron. Astrophys.*, **15**, 505.
G 5: Bahcall, N. A. 1980, *Astrophys. J. (Letters)*, **238**, L117.
G 6: Bardeen, J. M. 1975, in *Dynamics of Stellar Systems, IAU Symp.*, No. 69, ed. A. Hayli (D. Reidel Publ. Co., Dordrecht), p. 297.
G 7: Bautz, L. P., and Morgan, W. W. 1970, *Astrophys. J. (Letters)*, **162**, L149.
G 8: Corwin Jr., H. G., de Vaucouleurs, A., and de Vaucouleurs, G. 1982, *Astron. J.*, **87**, 47.
G 9: Curtis, H. D. 1919, quoted in Hubble (G24).
G10: de Vaucouleurs, G. 1948, *Ann. d'Astrophys.*, **11**, 247.
G11: de Vaucouleurs, G. 1956, *Mem. Commonwealth Obs.*, No. 13.
G12: de Vaucouleurs, G. 1959, in *Handbuch der Physik*, Bd. 53, ed. S. Flügge (Springer-Verlag, Berlin), p. 275.
G13: de Vaucouleurs, G. 1962, in *Problems of Extra-Galactic Research, IAU Symp.*, No. 15, ed. G. C. McVittie (The MacMillan Co., New York), p. 3.
G14: de Vaucouleurs, G. 1963, *Astrophys. J. Suppl.*, **8**, 31.
G15: de Vaucouleurs, G. 1974, in *The Formation and Dynamics of Galaxies, IAU Symp.* No. 58, ed. J. R. Shakeshaft (D. Reidel Publ. Co., Dordrecht), p. 335.
G16: de Vaucouleurs, G. 1977, *Astrophys. J. Suppl.*, **33**, 211.
G17: de Vaucouleurs, G. 1980, *Proc. Conf. on Photometry, Kinematics and Dynamics of Galaxies*, ed. D. S. Evans (University of Texas, Austin), p. 1.
G18: Freeman, K. C. 1970, *Astrophys. J.*, **160**, 811
G19: Haro, G. 1956, *Bol. Tonantzintla y Tacubaya*, No. 14, 8.
G20: Hausman, M. A., and Ostriker, J. P. 1978, *Astrophys. J.*, **224**, 220.
G21: Herschel, W. 1786, *Philosoph. Trans.*, **76**.
G22: Herschel, J. 1864, *Philosoph. Trans.*, **154**, 1.
G23: Hodge, P. W. 1966, in *The Physics and Astronomy of Galaxies and Cosmology* (McGraw-Hill Book Co., New York).
G24: Hubble, E. 1922, *Astrophys. J.*, **56**, 162.
G25: Hubble, E. 1923, *Memorandum to the Commission on nebulae of the IAU.*
G26: Hubble, E. 1926, *Astrophys. J.*, **64**, 321.
G27: Hubble, E. 1936, *The Realm of the Nebulae* (Yale University Press, New Haven).
G28: Jeans, J. H. 1919, *Problems of Cosmogony and Stellar Dynamics*, (Cambridge University Press, Cambridge).
G29: Kalnajs, A. J. 1974, in *La Dynamique des Galaxies Spirales, Colloque Intern. CNRS*, No. 241, ed. L. Weliachew, p. 103.
G30: Kazaryan, M. A. and Kazaryan, É. S. 1982, *Astrofiz.*, **18**, 512.
G31: Kinman, T. D. 1983, *Monthly Notices Roy. Astron. Soc.*, **202**, 53.
G32: Kormendy, J. 1977, *Astrophys. J.*, **217**, 406.
G33: Kormendy, J. 1979, *Astrophys. J.*, **227**, 714.
G34: Kormendy, J. 1980, in *Proc. ESO workshop on Two Dimensional Photometry*, ed. P. Crane and K. Kjär, (European Southern Observatory), p. 191.
G35: Markarian, B. E., Lipovetskij, V. A., and Stepanian, Dzh. A. 1981, *Astrofiz.*, **17**, 619.
G36: Matthews, T. A., Morgan, W. W., and Schmidt, M. 1964, *Astrophys. J.*, **140**, 35.
G37: McAlpine, G. M., and Williams. G. A. 1981, *Astrophys. J. Suppl.*, **45**, 113.
G38: Morgan, W. W. 1958, *Publ. Astron. Soc. Pacific*, **70**, 364.
G39: Morgan, W. W. 1959, *Publ. Astron. Soc. Pacific*, **71**, 394.
G40: Morgan, W. W., and Mayall, N. U. 1957, *Publ. Astron. Soc. Pacific*, **69**, 291.
G41: Morgan, W. W., Kayser, S., and White, R. A. 1975, *Astrophys. J.*, **199**, 545.
G42: Oemler, A. 1974, *Astrophys. J.*, **194**, 1.
G43: Osmer, P. S., and Smith, M. G. 1980, *Astrophys. J. Suppl.*, **42**, 333.
G44: Pesch, P., and Sanduleak, N. 1983, *Astrophys. J. Suppl.*, **51**, 171.
G45: Rood, H. J., and Sastry, G. N. 1971, *Publ. Astron. Soc. Pacific*, **83**, 313.
G46: Sandage, A. 1975, in *Galaxies and the Universe, Stars and Stellar Systems*, Vol. 9, ed. A. Sandage, M. Sandage, and J. Kristian (The University of Chicago Press, Chicago), p. 1.
G47: Savage, A., and Peterson, B. A. 1983, in *Proc. IAU Symp.* No. 104 (D. Reidel Publ. Co., Dordorecht), (in press).
G48: Schecter, P. 1976, *Astrophys. J.*, **203**, 297.
G49: Seyfert, C. K. 1943, *Astrophys. J.* **97**, 28.
G50: Shapley, H., and Ames, A. 1932, *Ann. Harvard Obs.*, **88**, No. 2.
G51: Schmidt, M., and Green, R. F. 1983, *Astrophys. J.*, **269**, 352.
G52: Simien, F., Athanassoula, E., Pellet, A., Monnet, G., Maucherat, A., and Courtès, G. 1978, *Astron. Astrophys.*, **67**, 73.
G53: Struble, M. F., and Rood, H. J. 1982, *Astron. J.*, **87**, 7.
G54: Takase, B., Ishida, K., Shimizu, M., Maehara, H., Hamajima, K., Noguchi, T., and Ohashi, M. 1977, *Annals Tokyo Astron. Obs.*, 2nd Ser., **16**, 74.
G55: Tammann, G. A., Yahil, A., and Sandage, A. 1979, *Astrophys. J.*, **234**, 775.
G56: Tsikoudi, V. 1979, *Astrophys. J.*, **234**, 842.
G57: Tsikoudi, V. 1980, *Astrophys. J. Suppl.*, **43**, 365.
G58: Turner, E. L., and Gott, J. R. 1976, *Astrophys. J. Suppl.*, **32**, 409.
G59: van den Bergh, S. 1960a, *Astrophys. J.*, **131**, 215.
G60: van den Bergh, S. 1960b, *Astrophys. J.*, **131**, 558.
G61: van den Bergh, S. 1960c, *Publ. David Dunlap Obs.*, **2**, No. 6.
G62: van der Kruit, P. C., and Searle, L. 1981a, *Astron. Astrophys.*, **95**, 105.

G63: van der Kruit, P. C., and Searle, L. 1981b, *Astron. Astrophys.*, **95**, 116.
G64: van der Kruit, P. C., and Searle, L. 1982a, *Astron. Astrophys.*, **110**, 61.
G65: van der Kruit, P. C., and Searle, L. 1982b, *Astron. Astrophys.*, **110**, 79.
G66: Vorontsov-Velyaminov, B. A. 1959, *Atlas and Catalogue of Interacting Galaxies, Part* I (Moscow Univ., Moscow).
G67: Vorontsov-Velyaminov, B. A. 1977, *Astron. Astrophys. Suppl.*, **28**, 1.
G68: Wolf, M. 1908, *Publ. Astrophys. Inst. König*, (Heidelberg), **3**, No. 5.

## Our Publications Related to This Atlas

P 1: Aoki, S., Noguchi, M., and Iye, M. 1979, *Publ. Astron. Soc. Japan*, **31**, 737.
P 2: Hamabe, M., Kodaira, K., Okamura, S., and Takase, B. 1979, *Publ. Astron. Soc. Japan*, **31**, 431.
P 3: Hamabe, M., Kodaira, K., Okamura, S., and Takase, B. 1980, *Publ. Astron. Soc. Japan*, **32**, 197.
P 4: Hamabe, M. and Okamura, S. 1982, *Ann. Tokyo Astron. Obs.* 2nd Ser., **18,** 191.
P 5: Hamabe, M. 1982, *Publ. Astron. Soc. Japan*, **34**, 423.
P 6: Hamabe, M. and Wakamatsu, K. 1984, in preparation.
P 7: Iye, M., Hamabe, M., Watanabe, M., and Okamura, S. 1979, in *Photometry, Kinematics and Dynamics of Galaxies*, ed. D. S. Evans (University of Texas, Austin), p. 407.
P 8: Iye, M., Okamura, S., Watanabe, M., and Hamabe, M. 1982, *Astrophys. J.*, **256**, 103.
P 9: Kodaira, K., Okamura, S., and Watanabe, M. 1983, *Astrophys. J.* (*Letters*), **274**, L49.
P10: Noguchi, T., He, X.-T., Cannon, R. D., Smith, M. G., Okamura, S., and Takase, B. 1983, *Publ. Astron. Soc. Japan*, **35**, 337.
P11: Okamura, S. 1977, *Ann. Tokyo Astron. Obs.* 2nd Ser., **16**, 111.
P12: Okamura, S. 1978, *Publ. Astron. Soc. Japan*, **30**, 91.
P13: Okamura, S., Kanazawa, T., and Kodaira, K. 1976, *Publ. Astron. Soc. Japan*, **28**, 329.
P14: Okamura, S., Takase, B., and Kodaira, K. 1977, *Publ. Astron. Soc. Japan*, **29**, 567.
P15: Okamura, S., Kodaira, K., and Watanabe, M. 1984, *Astrophys. J.*, **280** (in press).
P16: Takase, B. 1980, *Publ. Astron. Soc. Japan.*, **32**, 605.
P17: Takase, B., Noguchi, T., and Maehara, H. 1983, *Ann. Tokyo Astron. Obs.* 2nd Ser., **19**, 440.
P18: Wakamatsu, K. and Hamabe, M. 1984, *Astrophys. J. Suppl.*, (in press).
P19: Watanabe, M., Kodaira, K., and Okamura, S. 1982, *Astrophys. J. Suppl.*, **50**, 1.
P20: Watanabe, M. 1983, *Ann. Tokyo Astron. Obs.*, 2nd Ser., **19**, 121.
P21: Watanabe, M., Kodaira, K., and Okamura, S. 1984, (in preparation).
P22: Yamagata, T., Maehara, H., Okamura, S., and Takase, B. 1984, in *Proc. 2nd Asian-Pacific Regional Meeting*, (Bandung), (in press).

## Representative Atlases and Catalogues

C 1: *An Atlas of Peculiar Galaxies:* Arp, H. 1966 (California Institute of Technology, Pasadena).
C 2: *The Hubble Atlas of Galaxies:* Sandage, A. 1961 (Carnegie Institution of Washington, Washington, D. C.).
C 3: *Galaxias Australes:* Sérsic, J. 1969 (Universidad Nacional de Cordoba, Cordoba).
C 4: *Reference Catalogue of Bright Galaxies:* de Vaucouleurs, G., and de Vaucouleurs, A. 1964 (The University of Texas Press, Austin).
C 5: *Second Reference Catalogue of Bright Galaxies:* de Vaucouleurs, G., de Vaucouleurs, A., and Corwin Jr., H. G. 1976 (University of Texas Press, Austin).
C 6: *A Revised Shapley-Ames Catalog of Bright Galaxies:* Sandage, A., and Tammann, G. A. 1981 (Carnegie Institution of Washington, Washington, D. C.).
C 7: *Morphological Catalogue of Galaxies,* Vol. 1: Vorontsov-Velyaminov, B. A. 1962 (Moscow)
C 8: *Morphological Catalogue of Galaxies,* Vol. 2: Vorontsov-Velyaminov, B. A. 1963 (Moscow)
C 9: *Morphological Catalogue of Galaxies,* Vol. 3: Vorontsov-Velyaminov, B. A. 1964 (Moscow)
C10: *Morphological Catalogue of Galaxies,* Vol. 4: Vorontsov-Velyaminov, B. A. 1968 (Moscow)
C11: *Cataloguë of Selected Compact Galaxies and of Post-Eruptive Galaxies:* Zwicky, F. 1971 (L. Speich, Zürich).
C12: *Catalogue of Galaxies and of Clusters of Galaxies,* Vol. I: Zwicky, F., Herzog, E., and Wild, P. 1961 (California Institute of Technology, Pasadena).
C13: *Catalogue of Galaxies and of Clusters of Galaxies,* Vol. II: Zwicky, F., and Herzog, E. 1963 (California Institute of Technology, Pasadena).
C14: *Catalogue of Galaxies and of Clusters of Galaxies,* Vol. III: Zwicky, F., and Herzog, E. 1966 (California Institute of Technology, Pasadena).
C15: *Catalogue of Galaxies and of Clusters of Galaxies,* Vol. IV: Zwicky, F., and Herzog, E. 1968 (California Institute of Technology, Pasadena).
C16: *Catalogue of Galaxies and of Clusters of Galaxies,* Vol. V: Zwicky, F., Karpowicz, M., and Kowal, C. T. 1965 (California Institute of Technology, Pasadena).
C17: *Catalogue of Galaxies and of Clusters of Galaxies,* Vol. VI: Zwicky, F., and Kowal, C. T. 1968 (California Institute of Technology, Pasadena).

# Part II. The Atlas

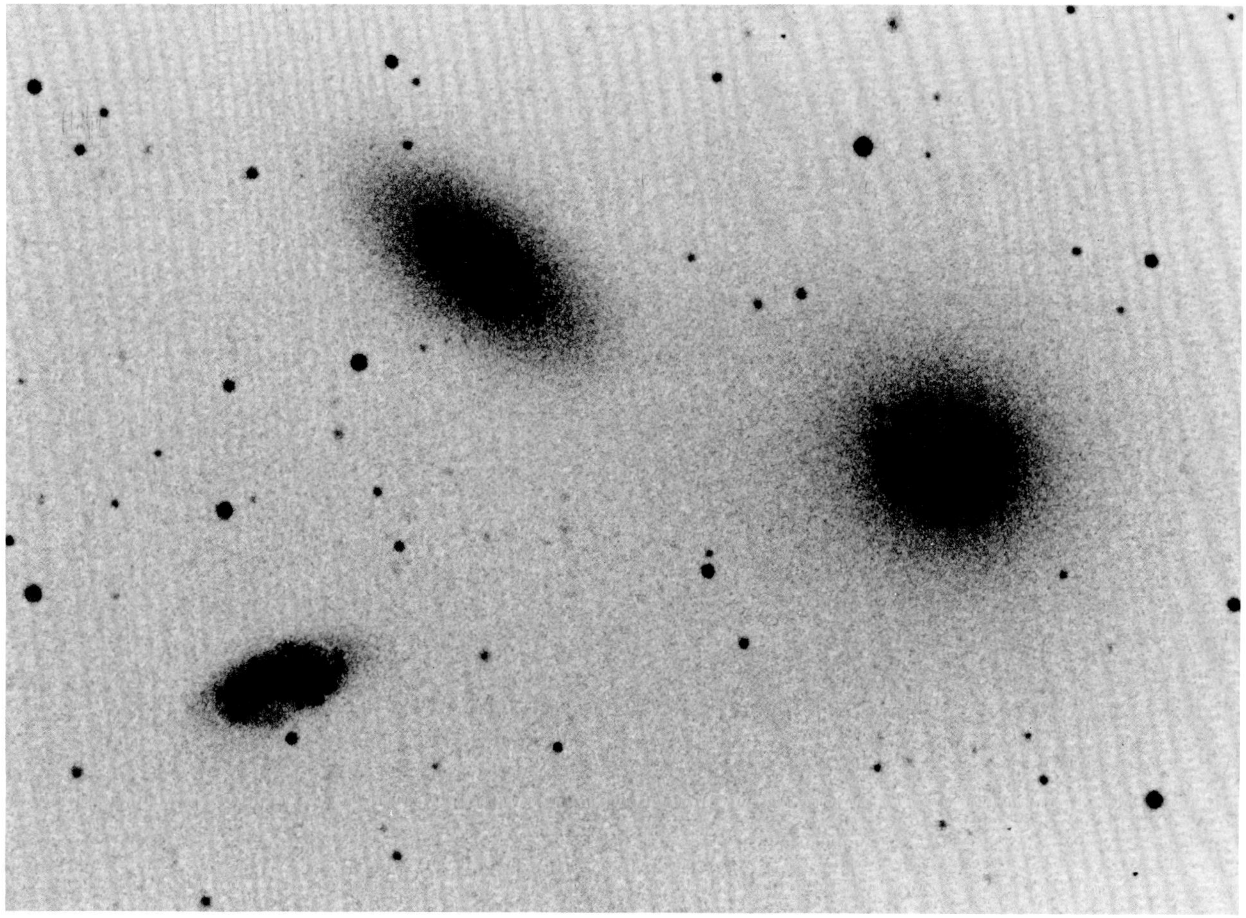

A1.1: NGC 3379 E1,
NGC 3384 SB(s)0$^{-}$:, NGC 3389 SA(s)c (B)

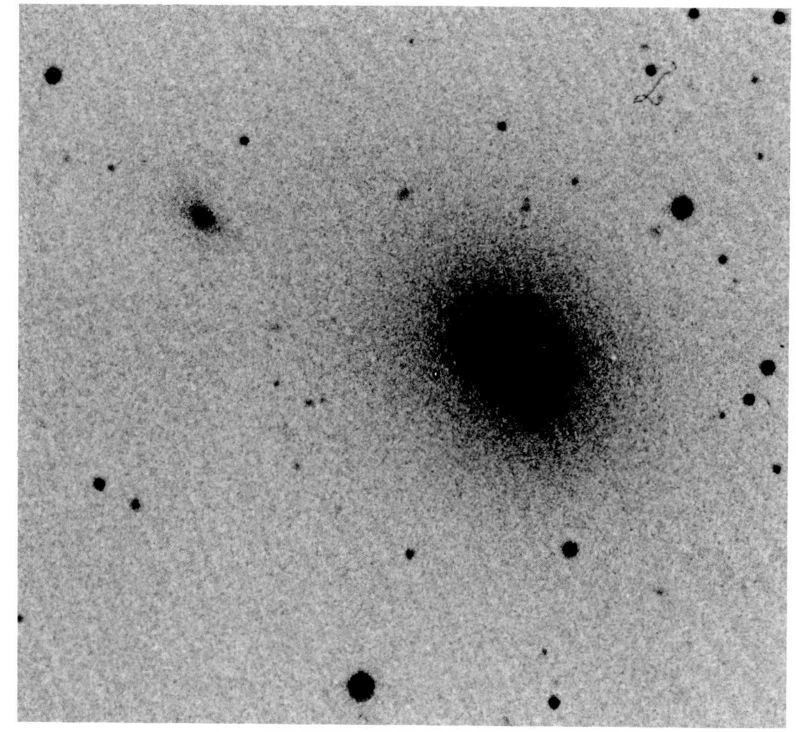

A1.2: NGC 4365 E3 (B)

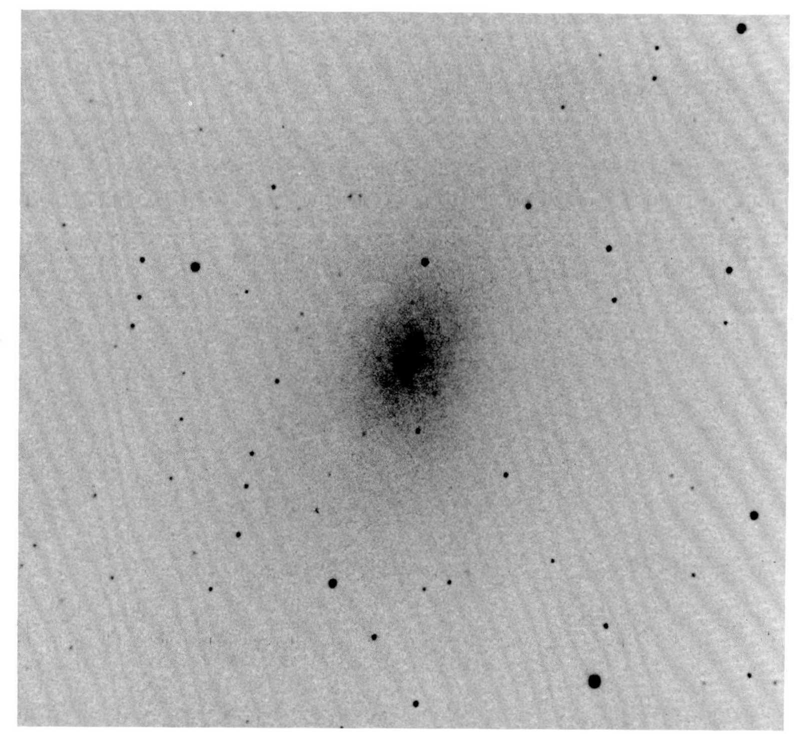

A1.3: NGC 205 E5P (V)

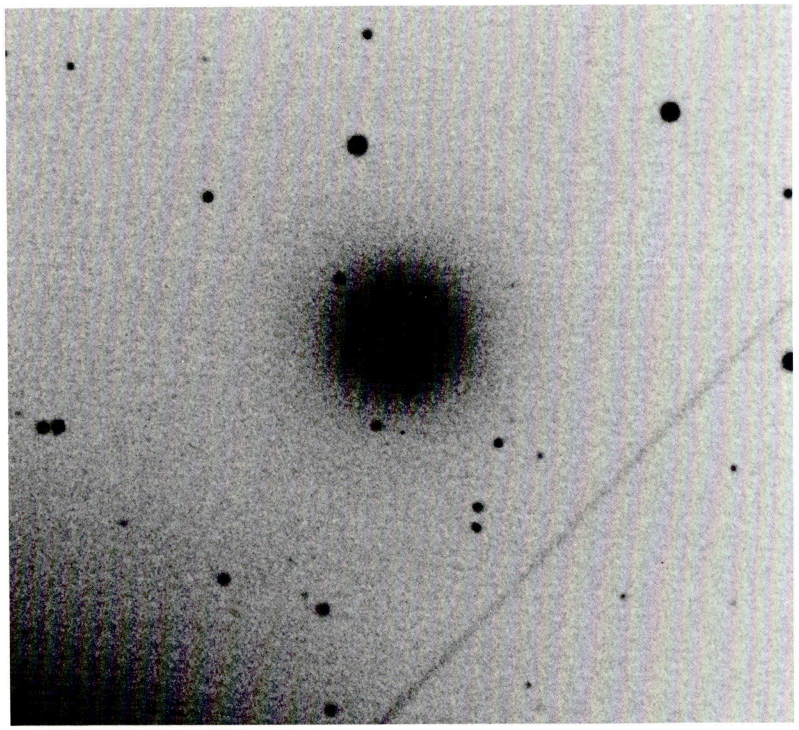

A1.4: NGC 404 SA(s)$0^-$: (B)

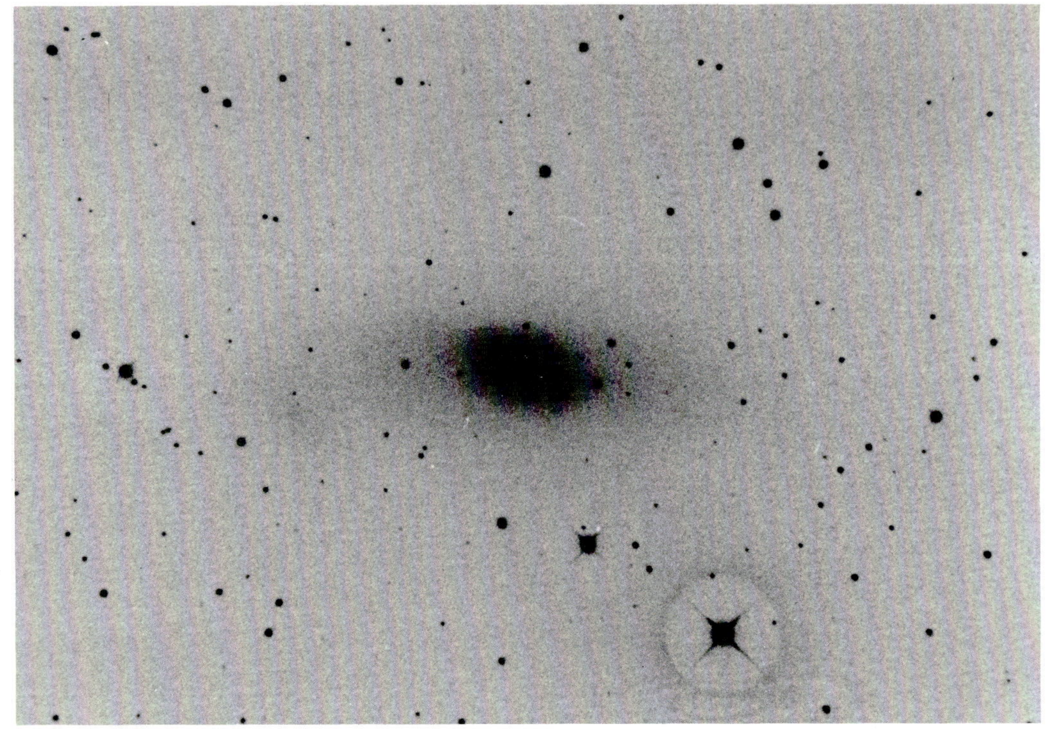

A1.5: NGC 1023 SB(rs)$0^-$ (B)

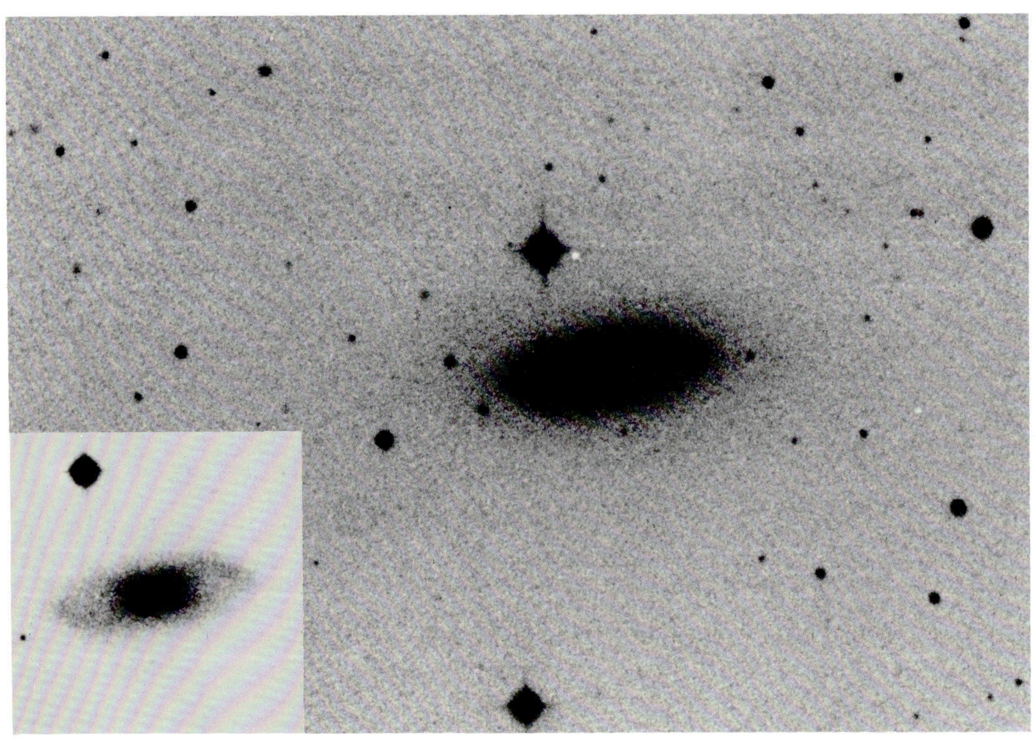

A1.6: NGC 4429 SA(r)$0^+$ (B)

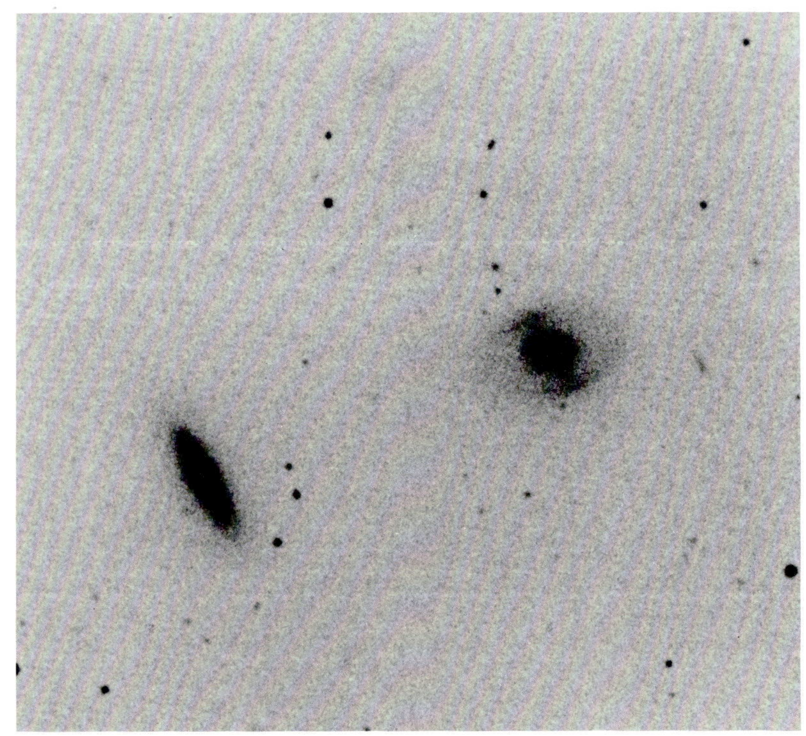

A1.7: NGC 4340 SB (r)$0^+$ (B)

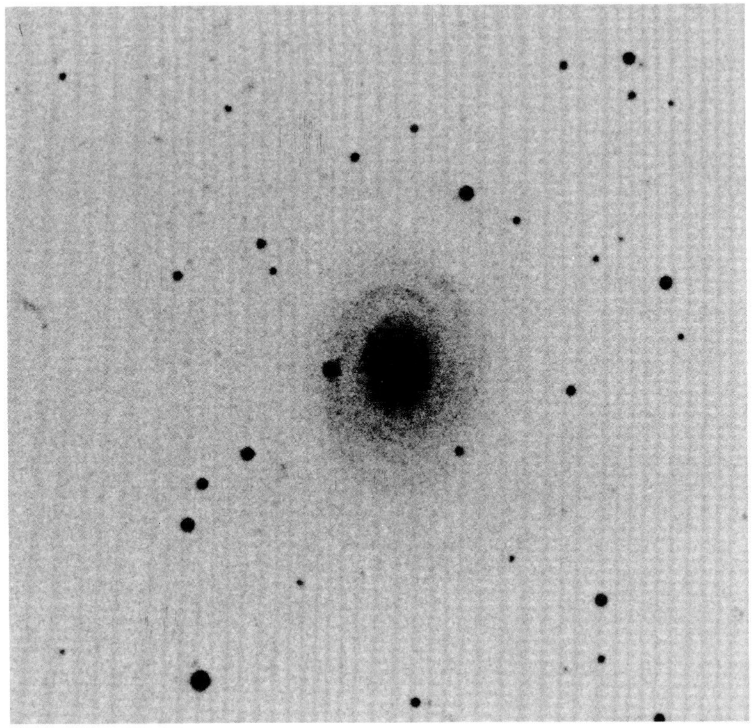

A1.8: NGC 2985 (R')SA(rs)ab (B)

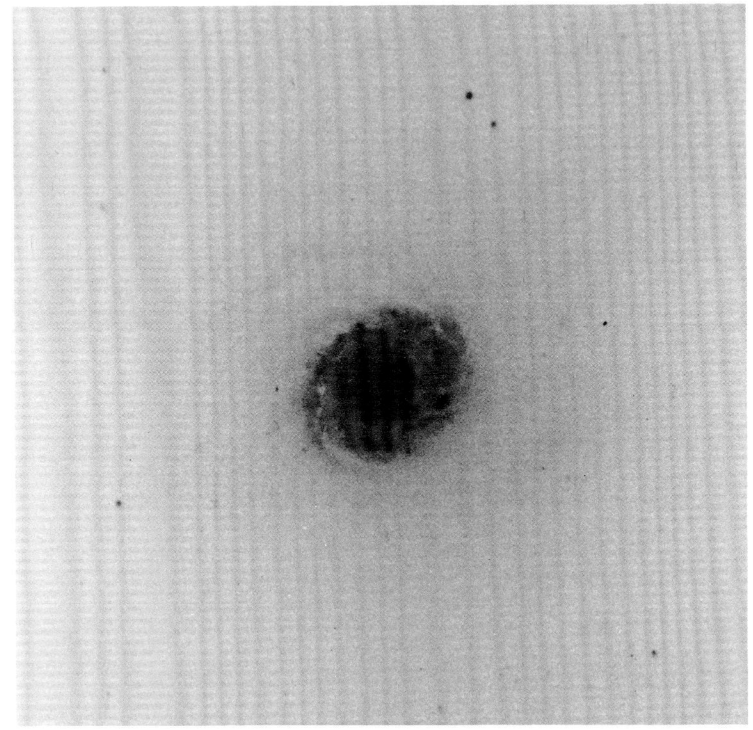

A1.9: NGC 4736 (M94) (R)SA(r)ab (B)

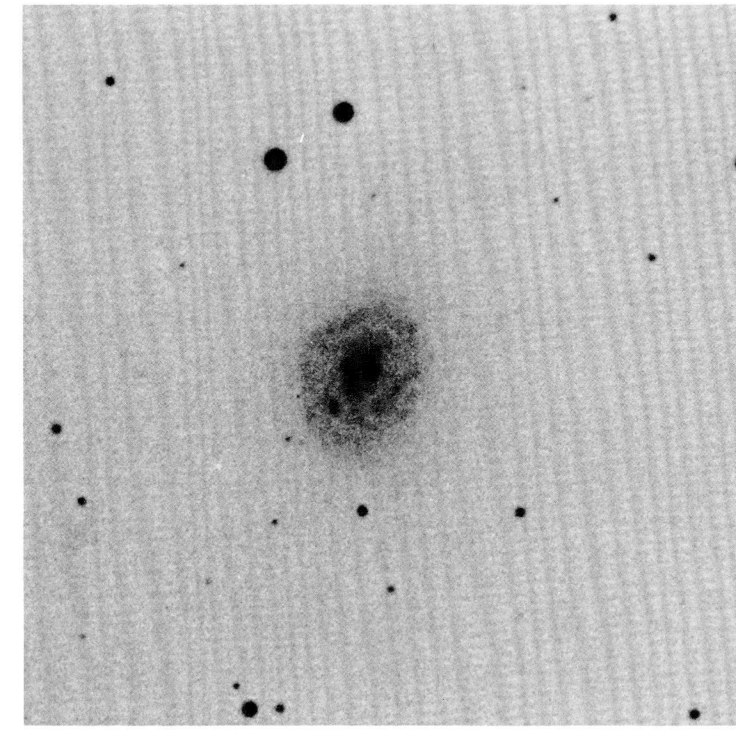

A1.10: NGC 4689 SA(rs)bc (B)

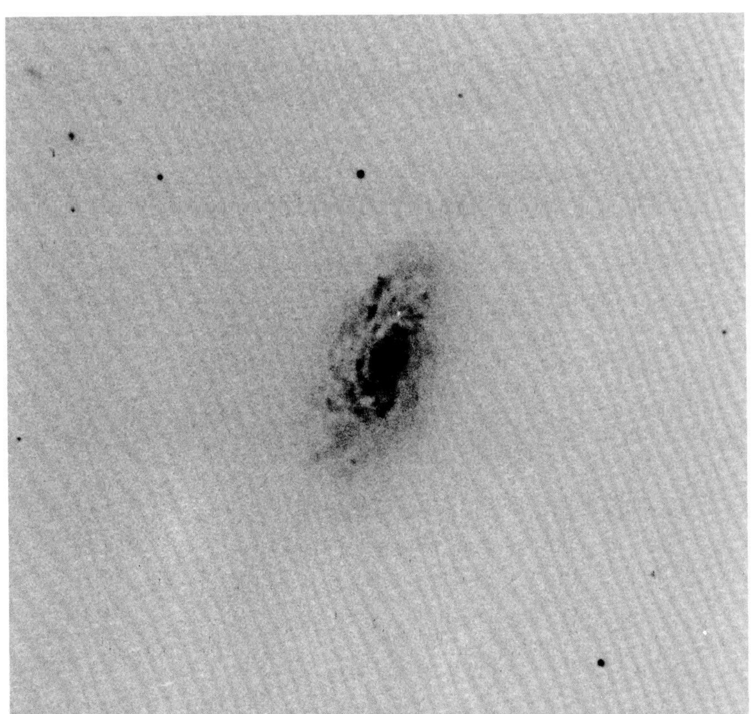

A1.11: NGC 4414 SA(rs)c? (B)

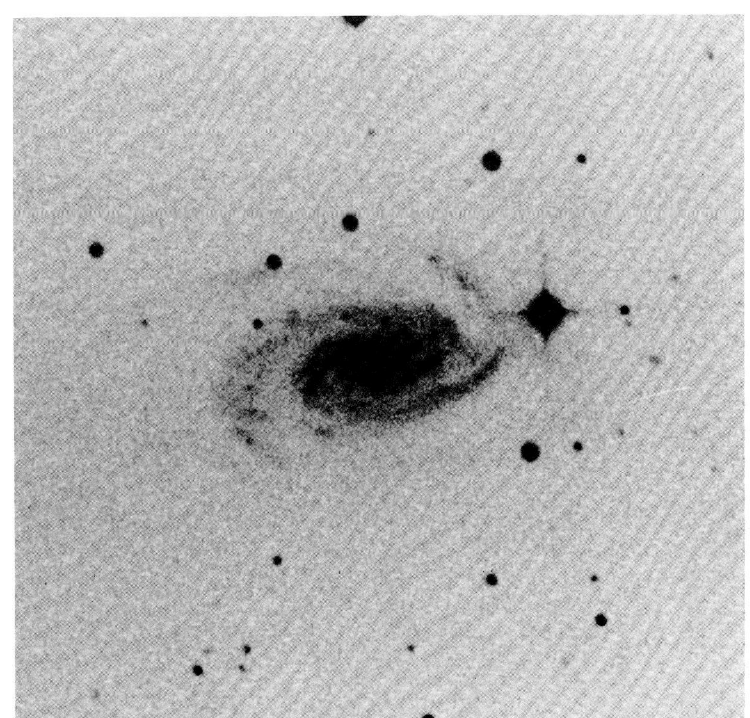

A1.12: NGC 3338 SA(s)c (B)

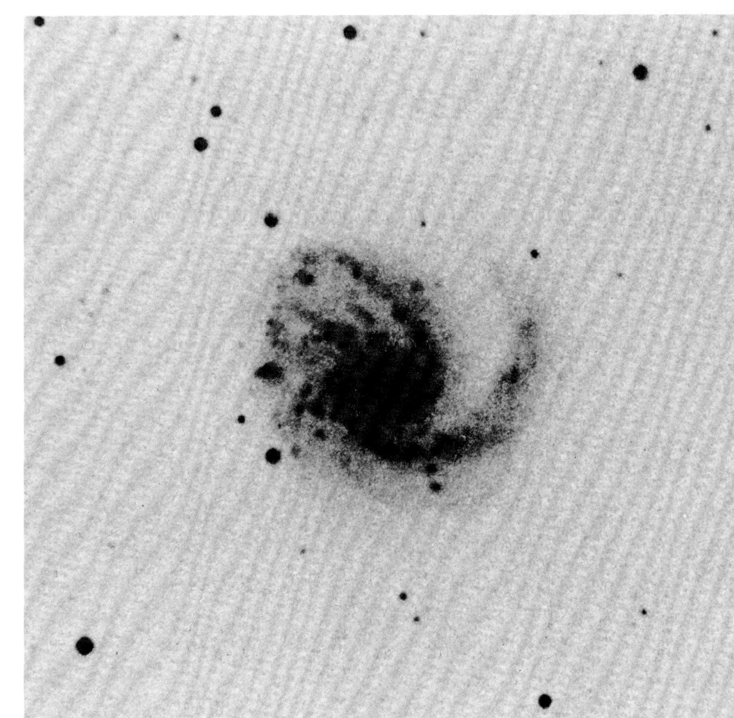

A1.13: NGC 4254 SA(s)c (B)

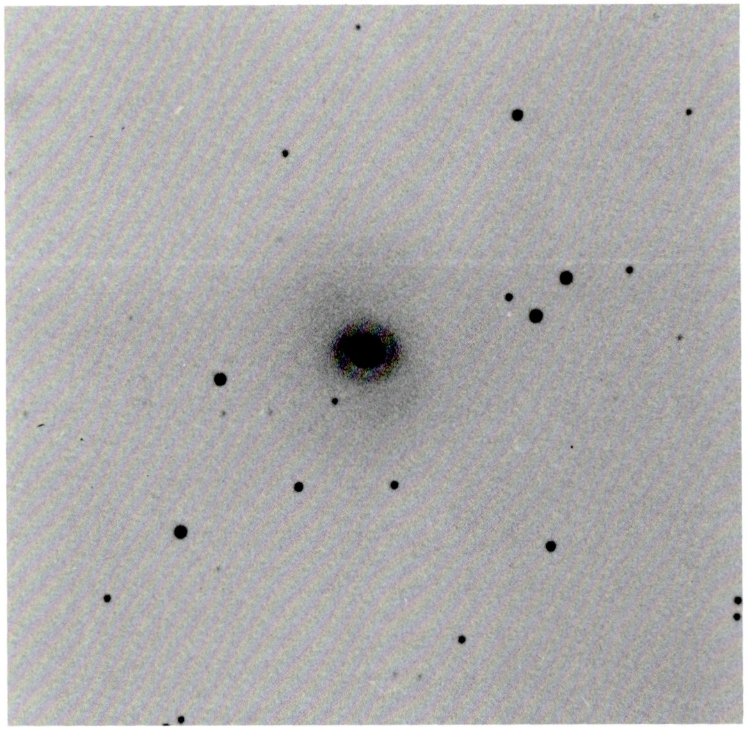

A1.14: NGC 2681 (R′)SAB(rs)0/a (B)

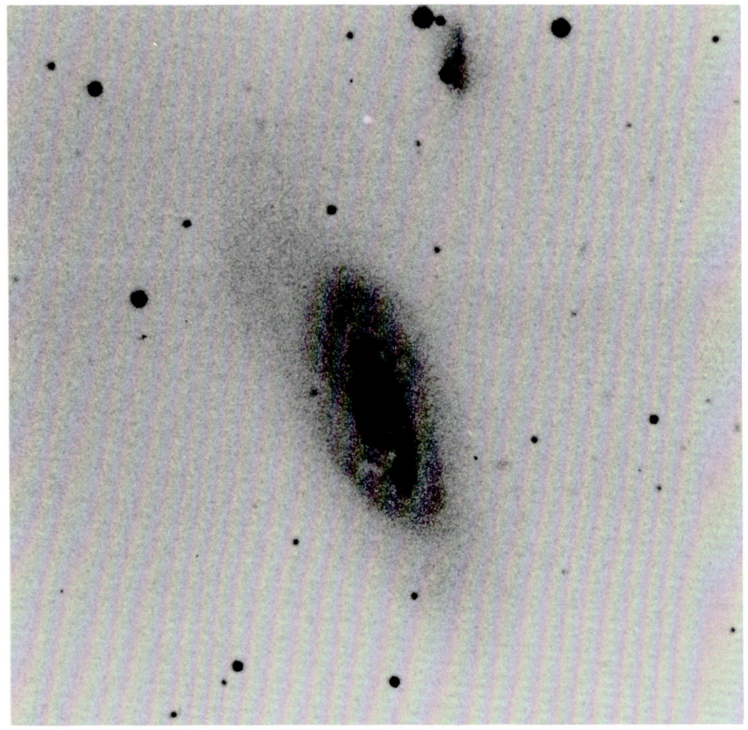

A1.15: NGC 4569 SAB(rs)ab, IC 3583 IBm (B)

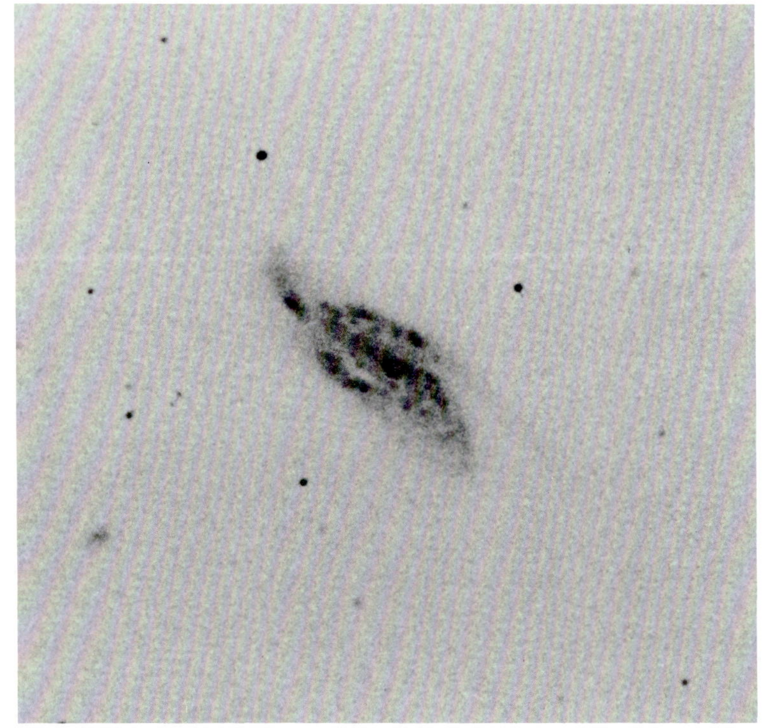

A1.16: NGC 4088 SAB(rs)bc (B)

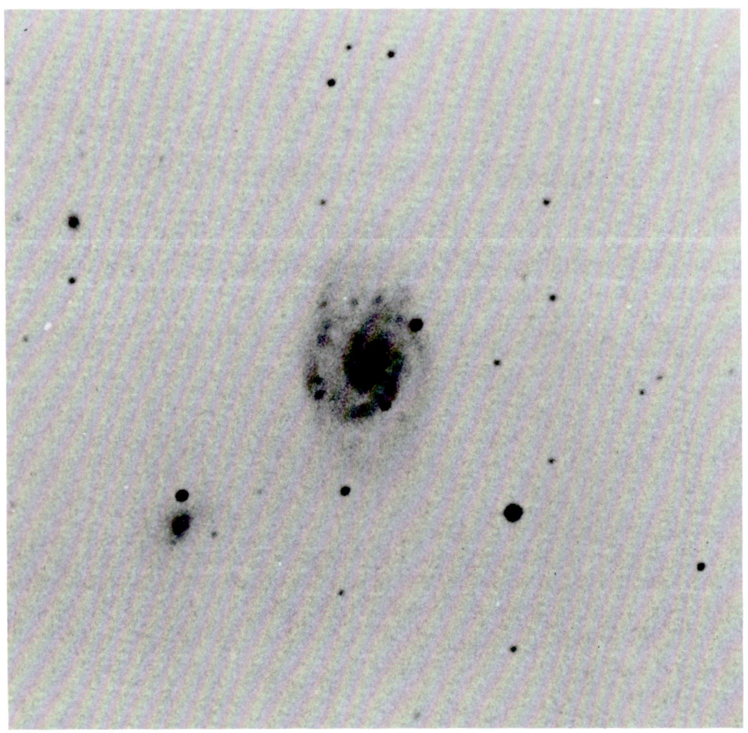

A1.17: NGC 3893 SAB(rs)c: (B)

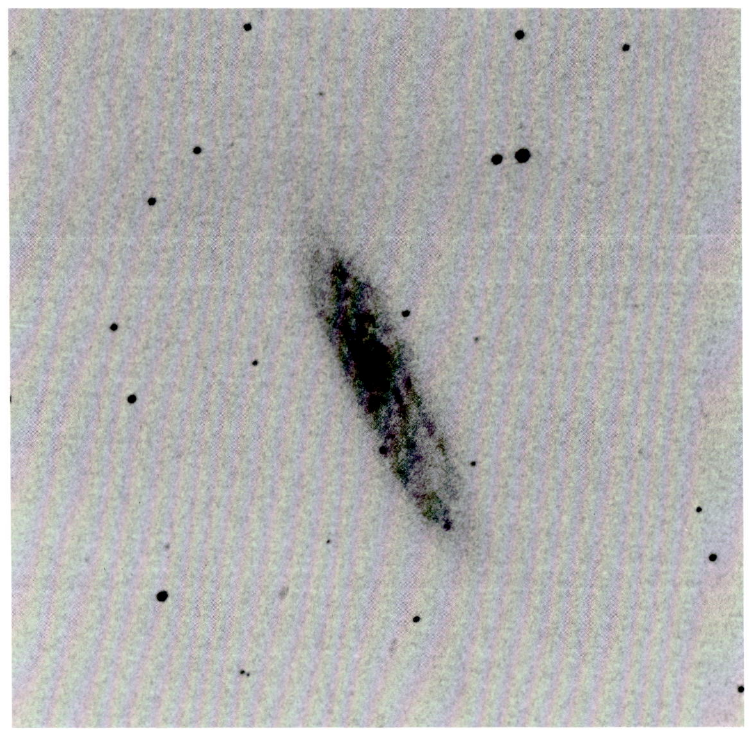

A1.18: NGC 4096 SAB(rs)c (B)

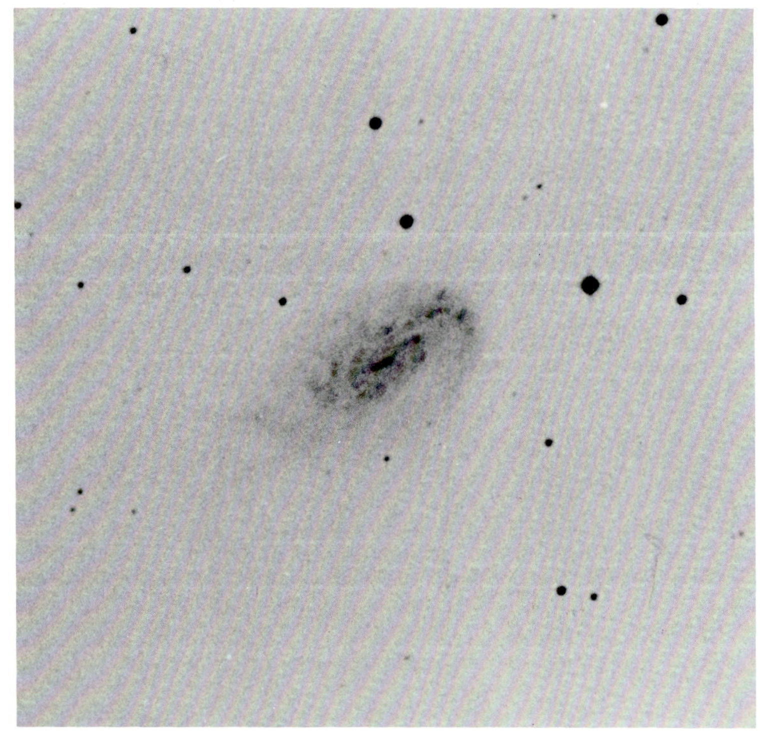

A1.19: NGC 4654 SAB(rs)cd (B)

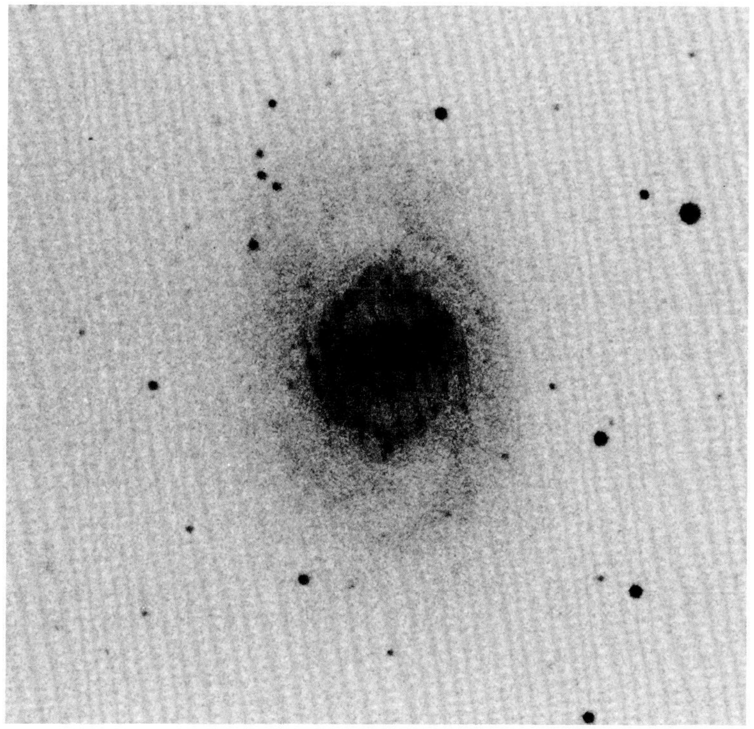

A1.20: NGC 3351 (M95) SB(r)b (B)

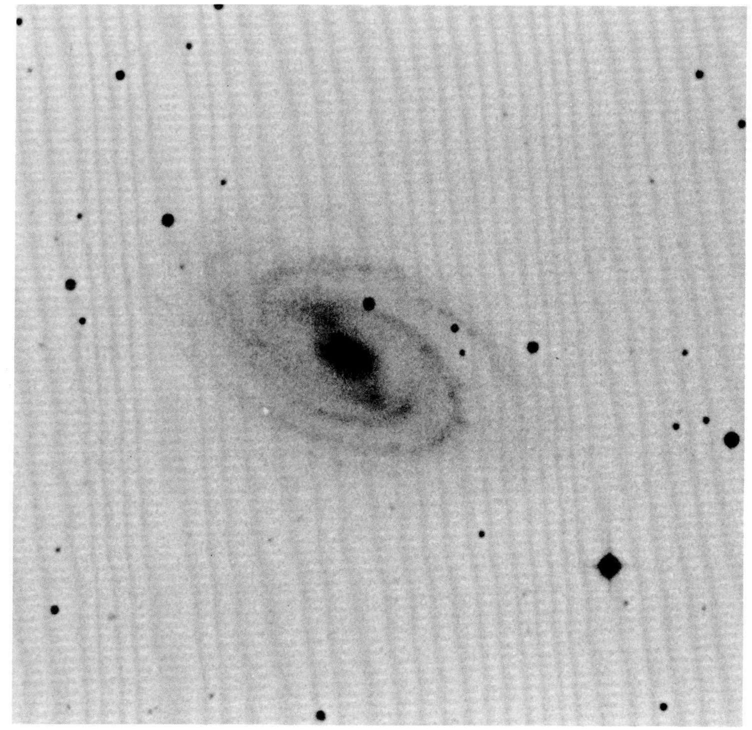

A1.21: NGC 3992 SB(rs)bc (V)

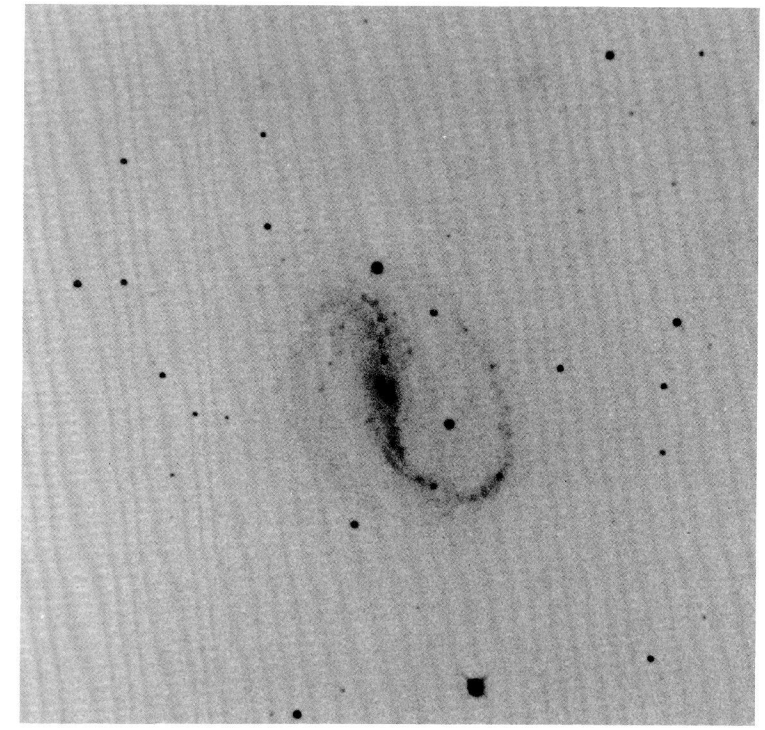

A1.22: NGC 7479 SB(s)c (B)

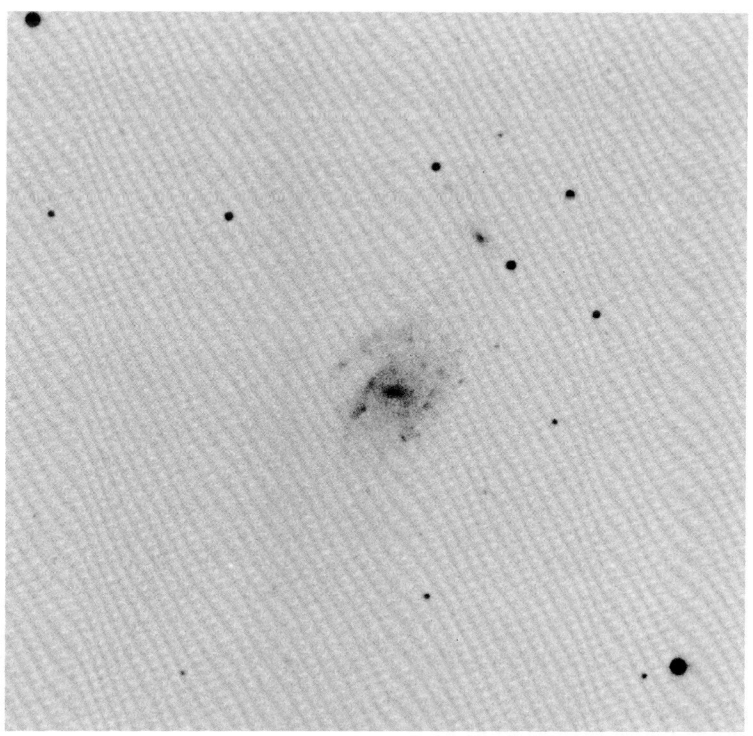

A1.23: NGC 4519 SB(rs)d (B)

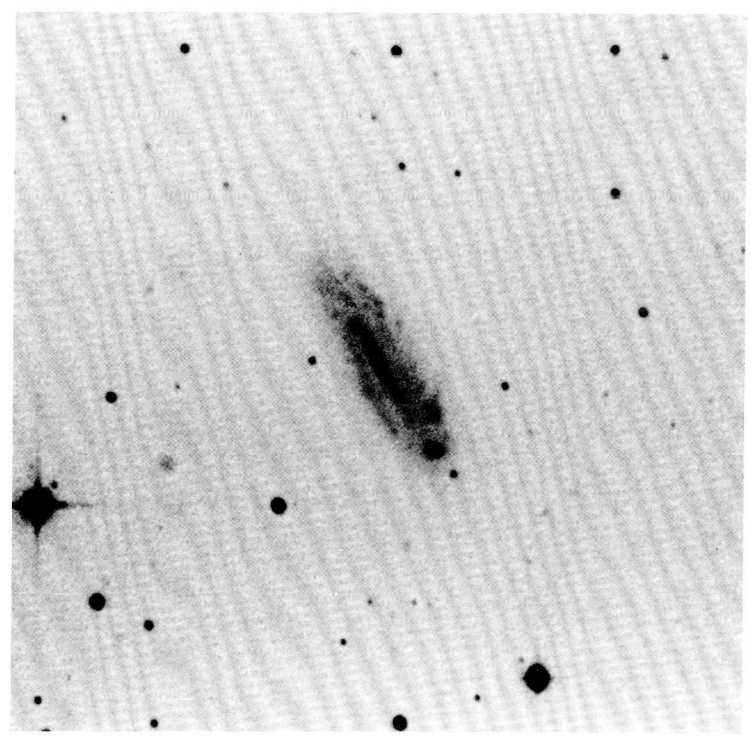

A1.24: NGC 4178 SB(rs)dm (B)

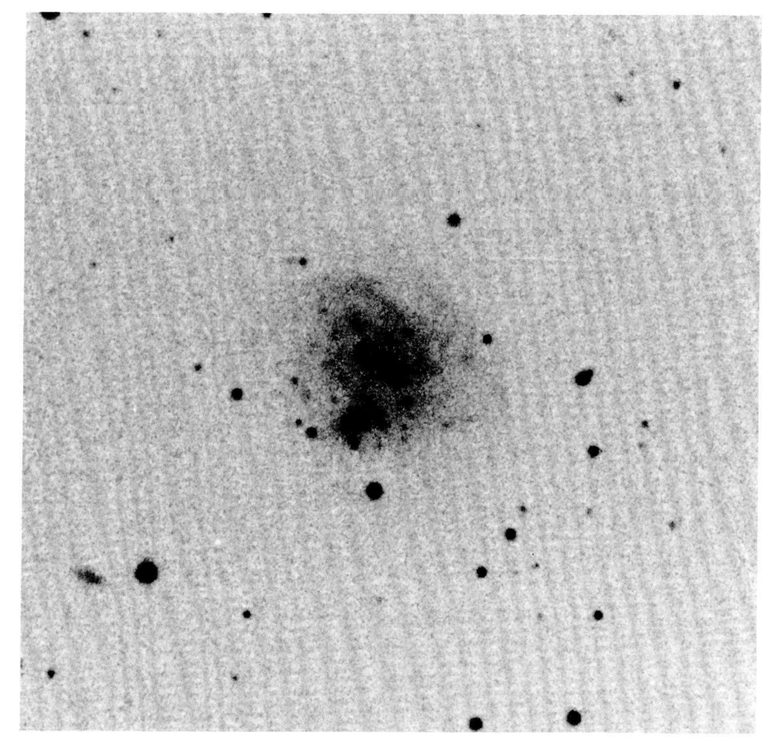

A1.25: NGC 4496A SB(rs)m
NGC 4496B IB(s)m: (B)

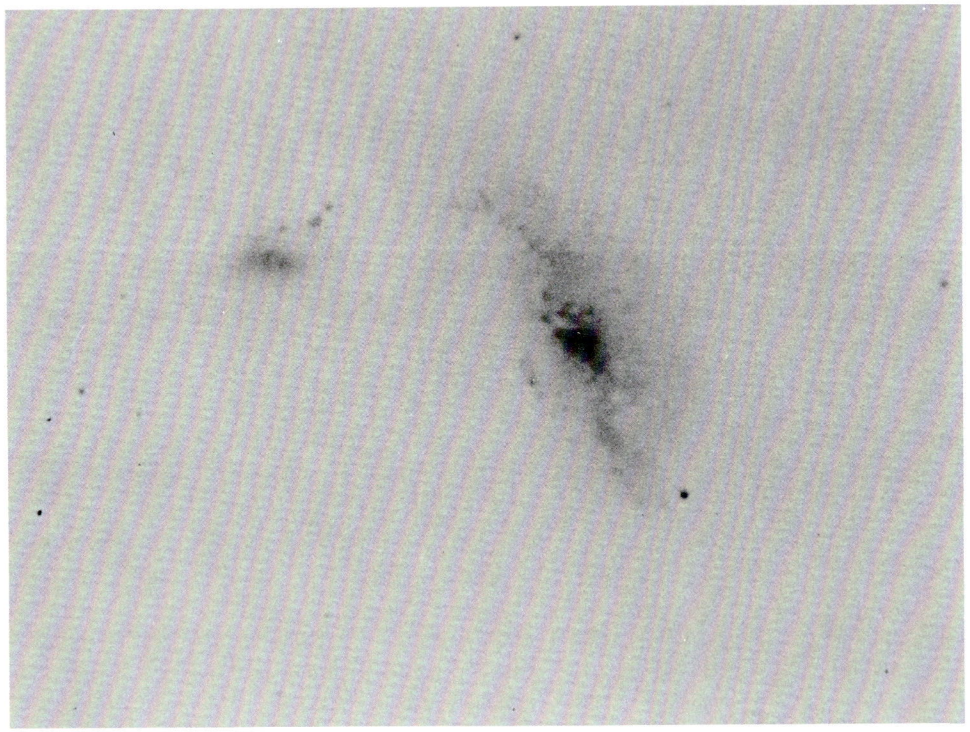

A1.26: NGC 4485 IB(s)mP (B)

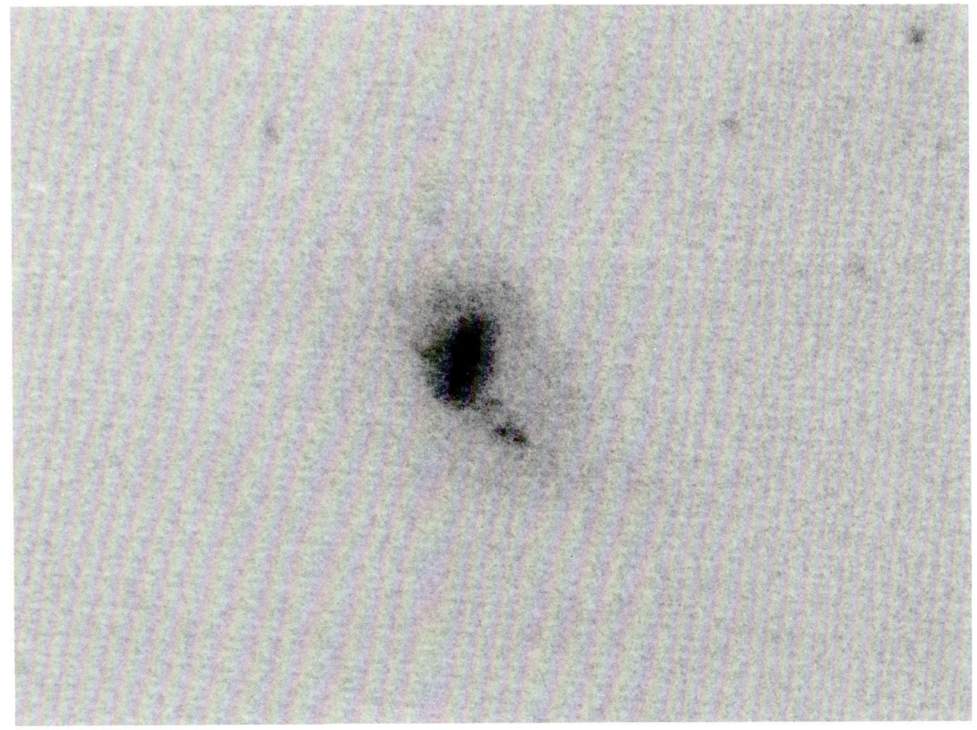

A1.27: IC 3476 IB(s)m: (V)

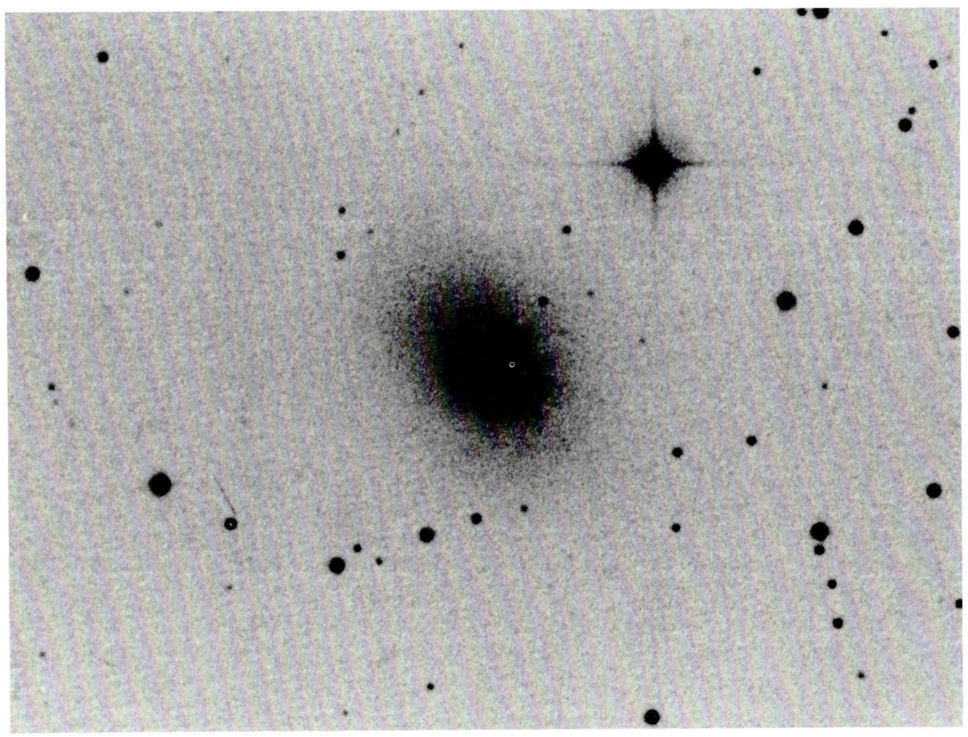

A1.28: NGC 3077 I0P (B)

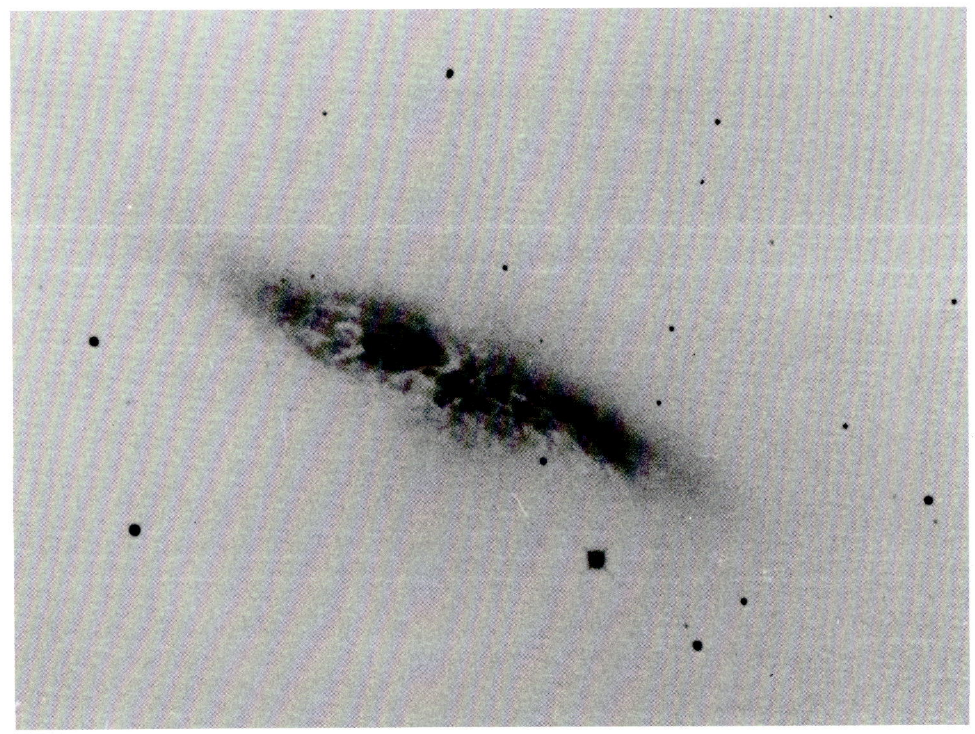

A1.29: NGC 3034 (M82) I0sp (B)

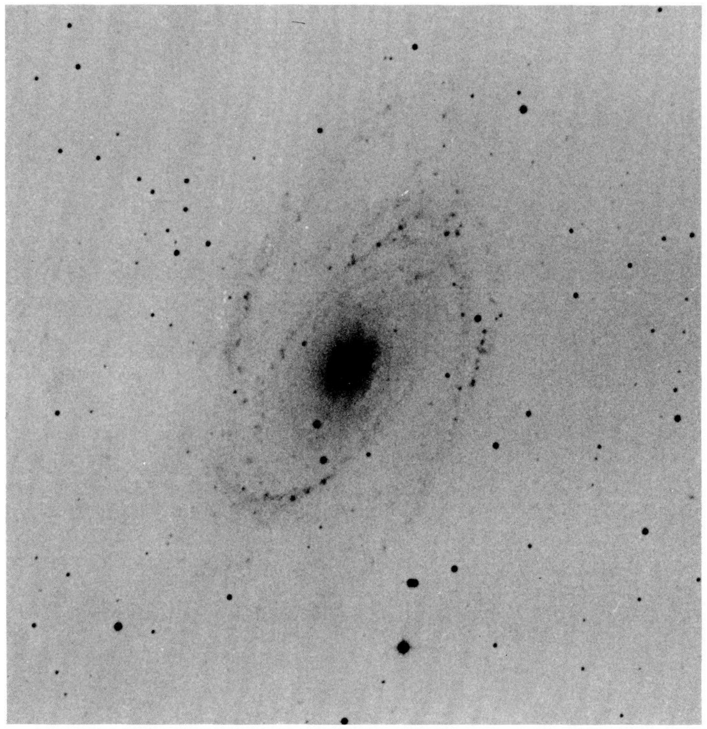

A2.1a: NGC 3031 (M81) SA(s)ab (U)

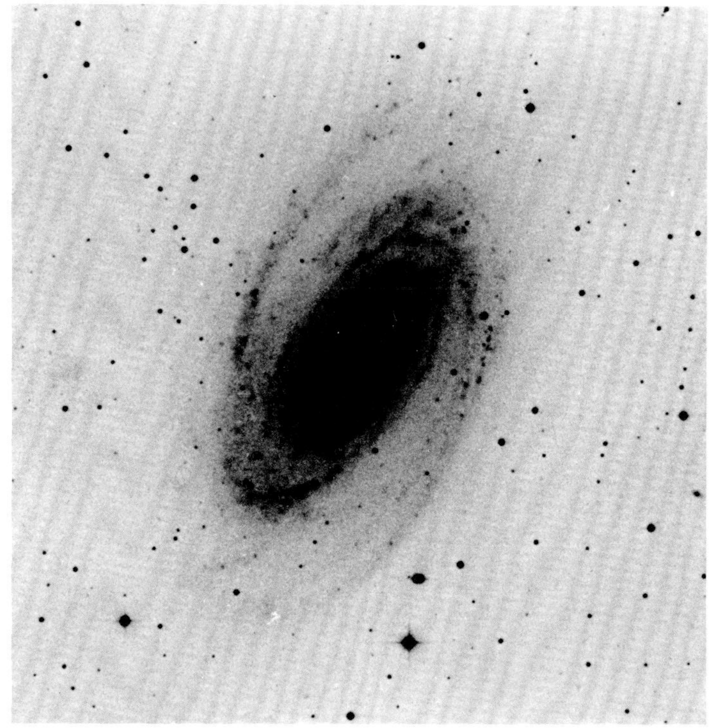

A2.2a: NGC 3031 (M81) SA(s)ab (B)

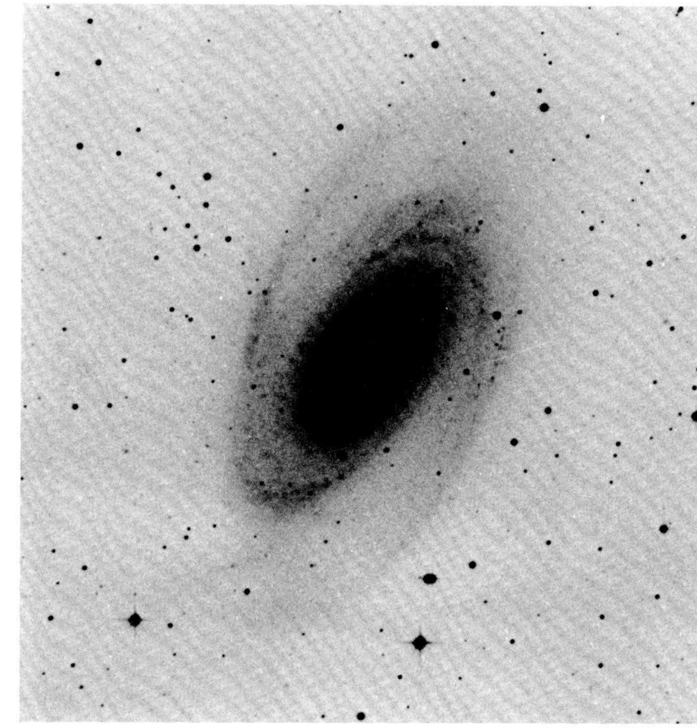

A2.3a: NGC 3031 (M81) SA(s)ab (V)

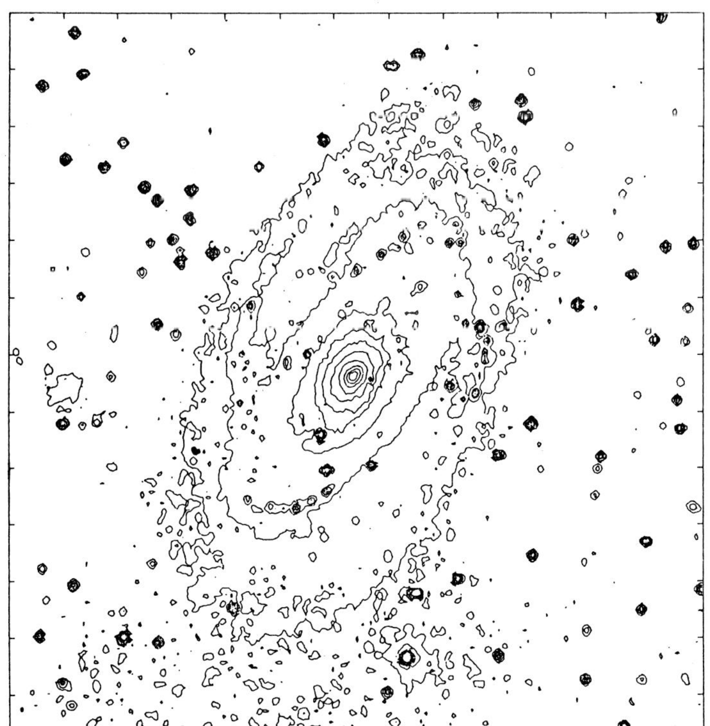

A2.1b: NGC 3031 (M81) SA(s)ab (U)

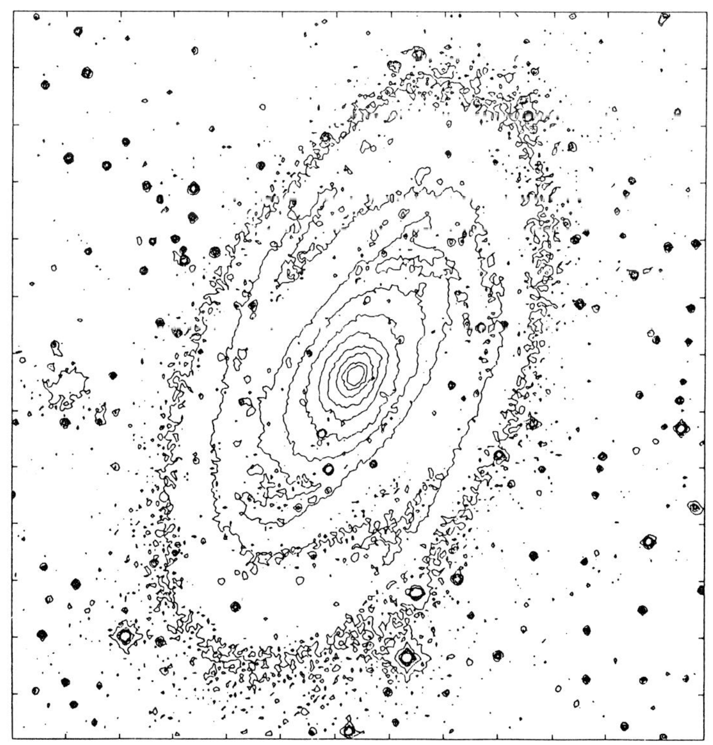

A2.2b: NGC 3031 (M81) SA(s)ab (B)

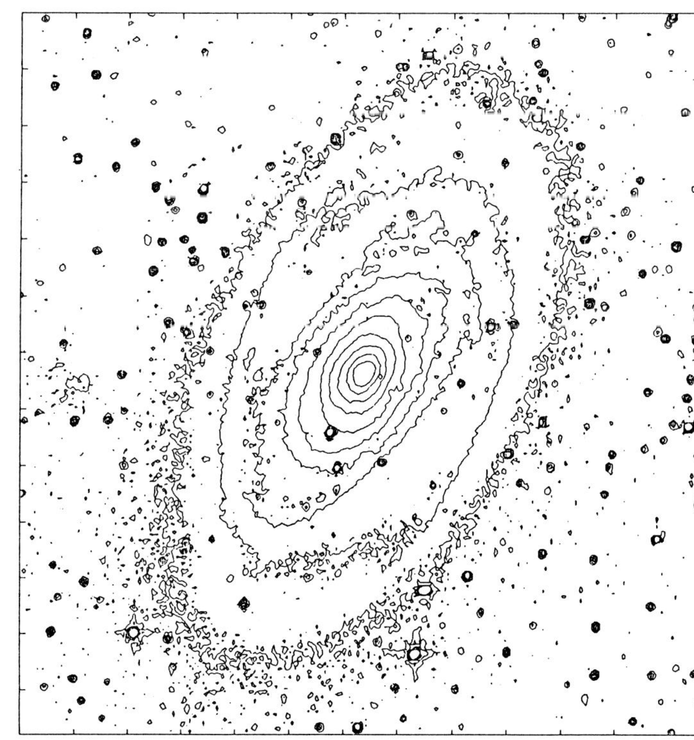

A2.3b: NGC 3031 (M81) SA(s)ab (V)

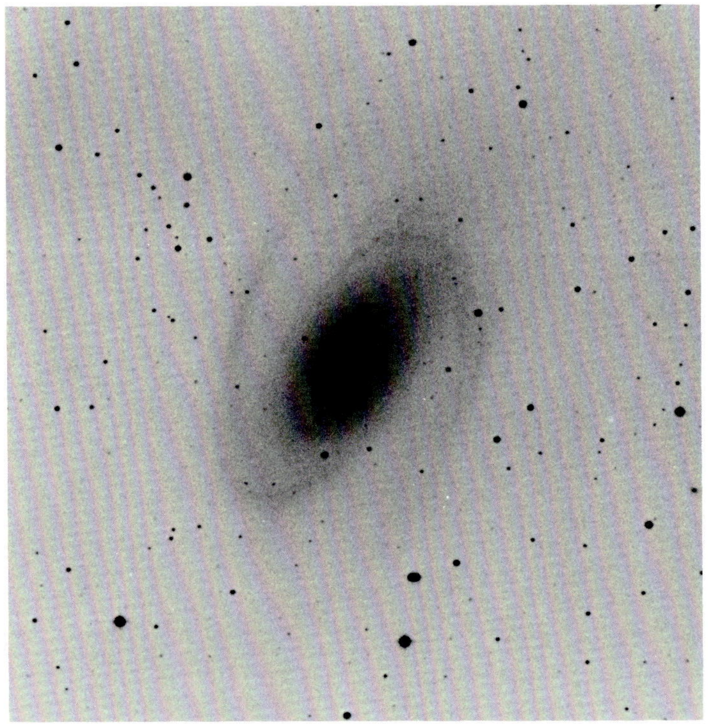

A2.4a: NGC 3031 (M81) SA(s)ab (I)

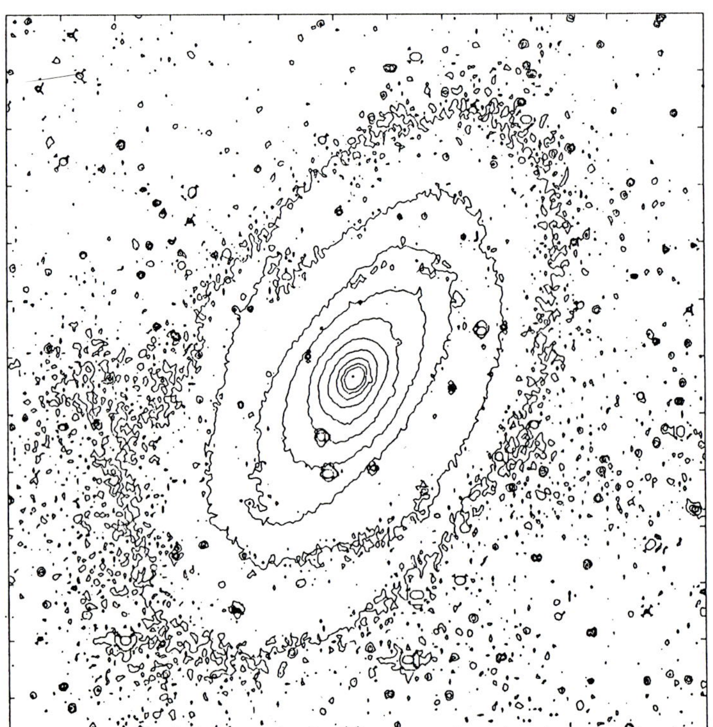

A2.4b: NGC 3031 (M81) SA(s)ab (I)

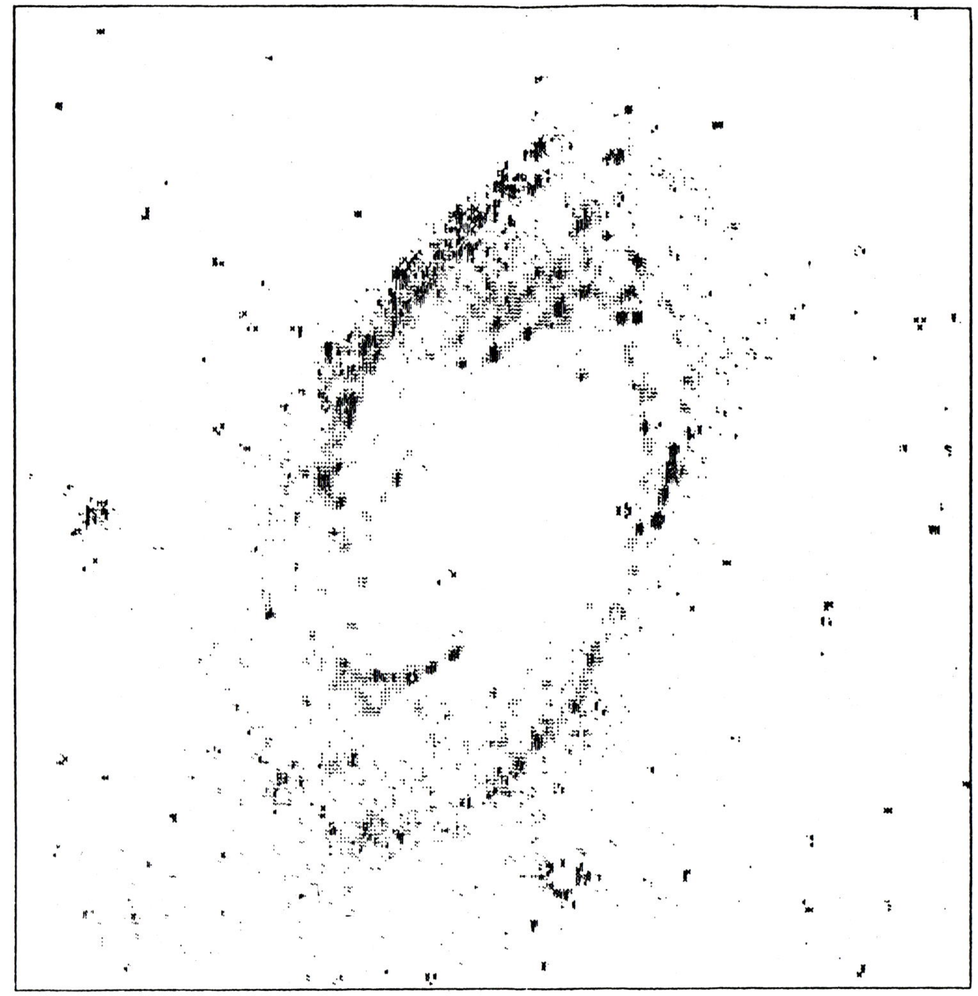

A2.5: NGC 3031 (M81) SA(s)ab (U-I)

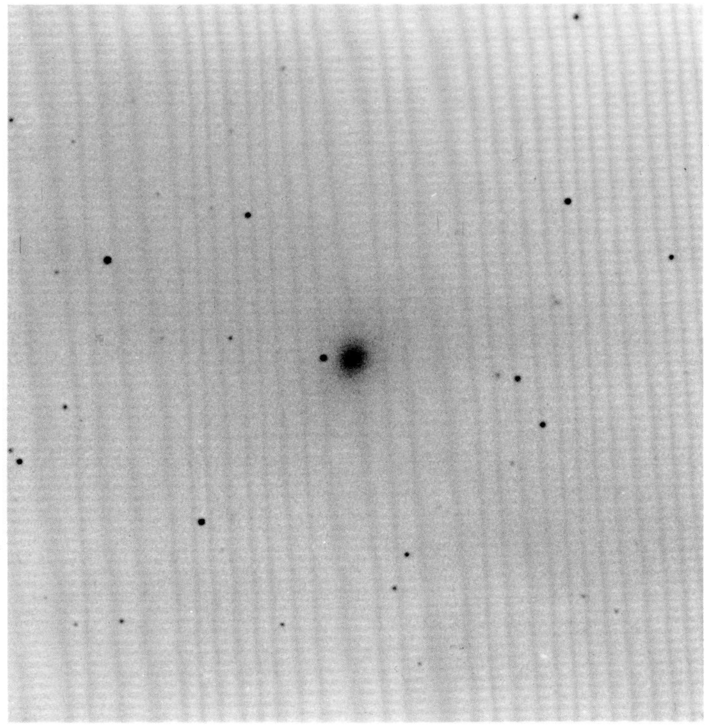

A2.6a: NGC 4472 E2 (U)

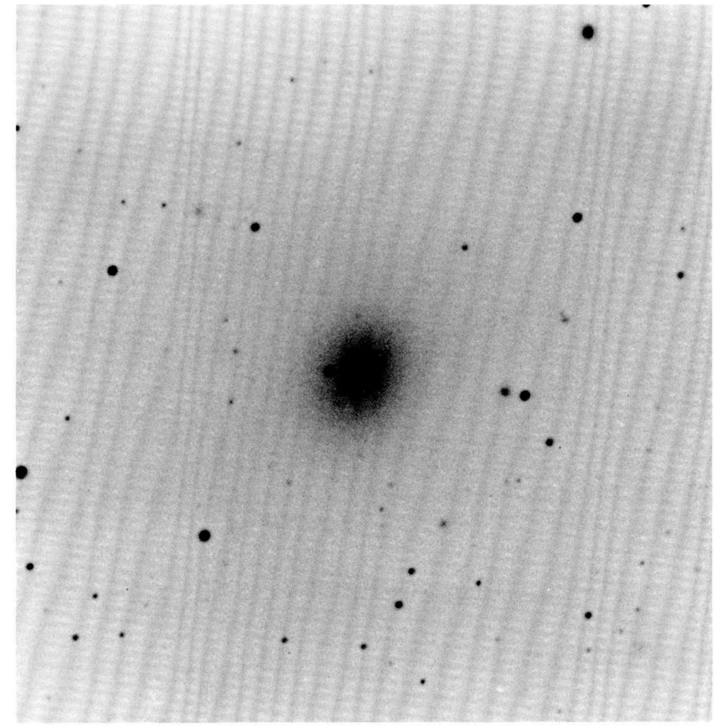

A2.7a: NGC 4472 E2 (I)

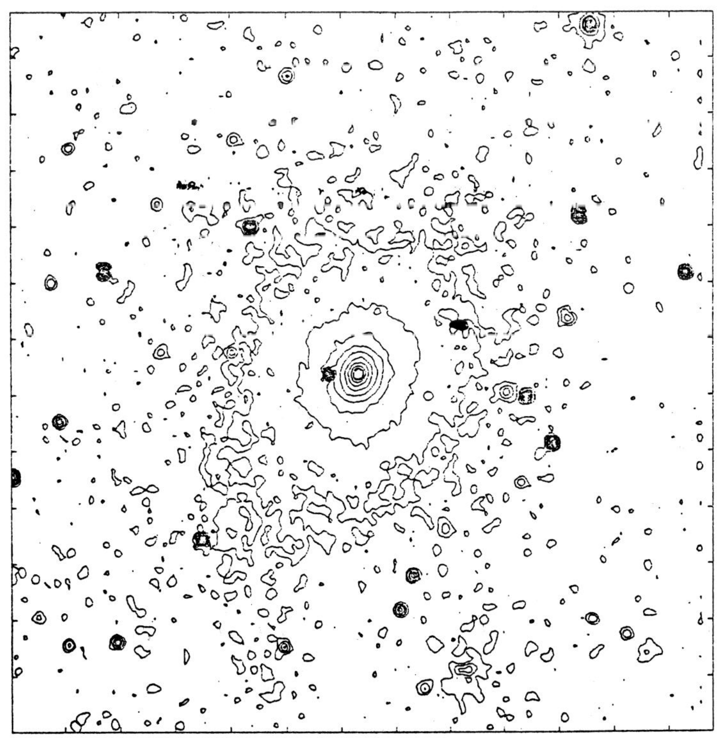

A2.6b: NGC 4472 E2 (U)

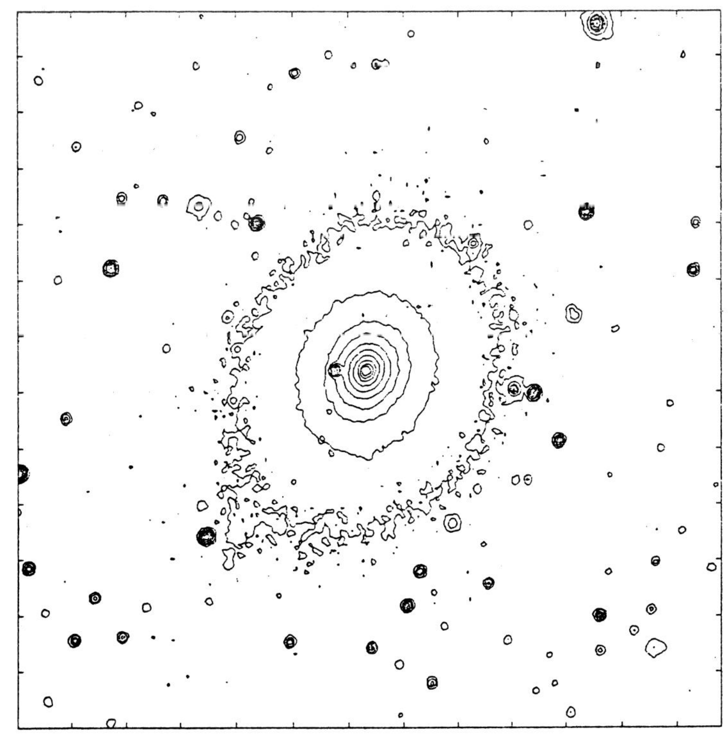

A2.7b: NGC 4472 E2 (I)

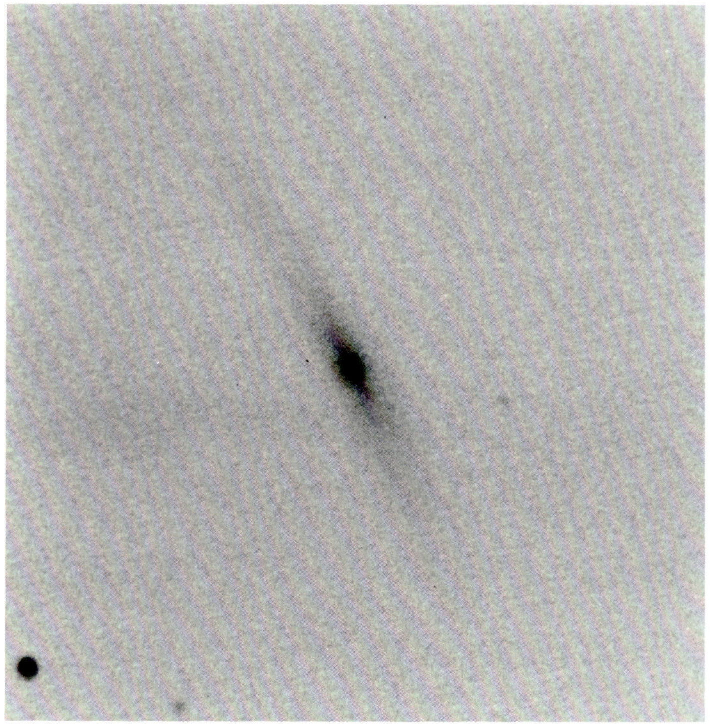

A2.8a: NGC 4350 SA0sp (U)

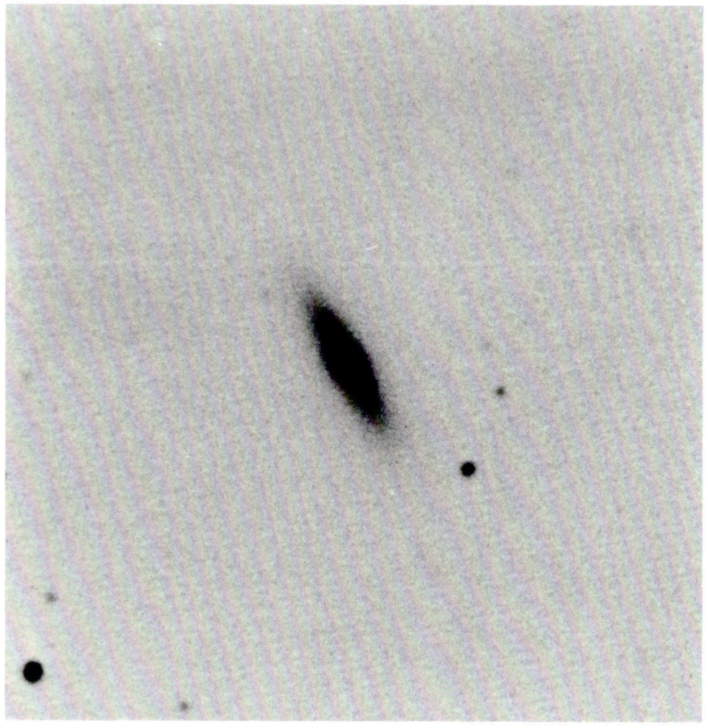

A2.9a: NGC 4350 SA0sp (I)

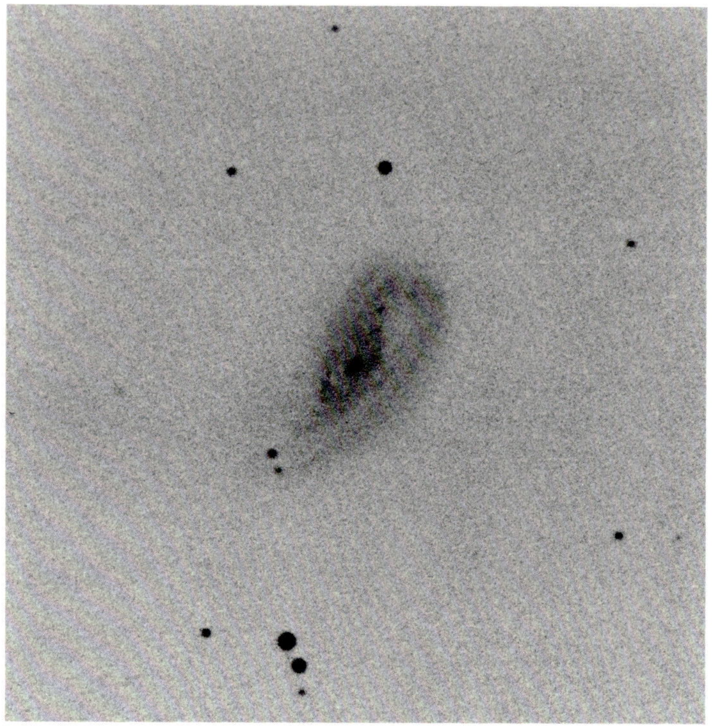

A2.10a: NGC 4501 SA(rs)b (U)

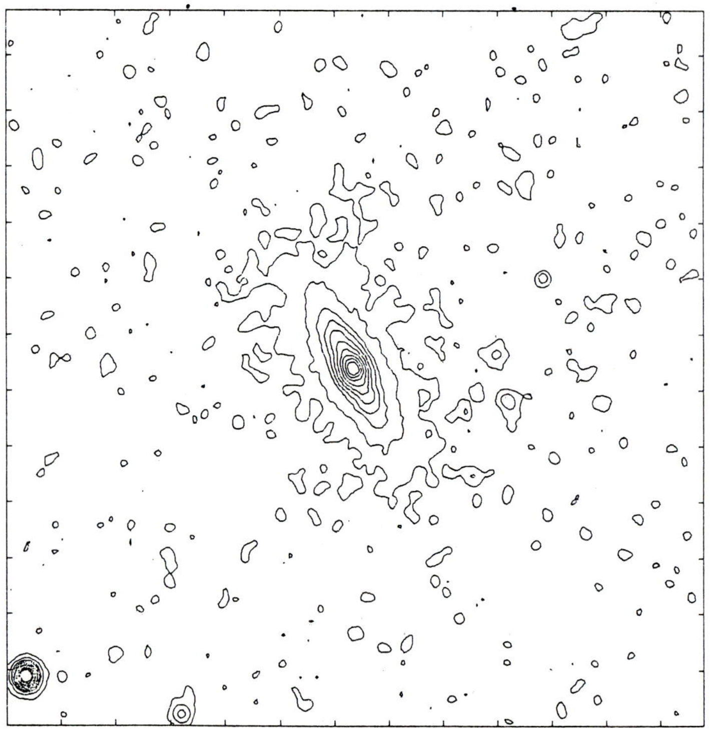

A2.8b: NGC 4350 SA0sp (U)

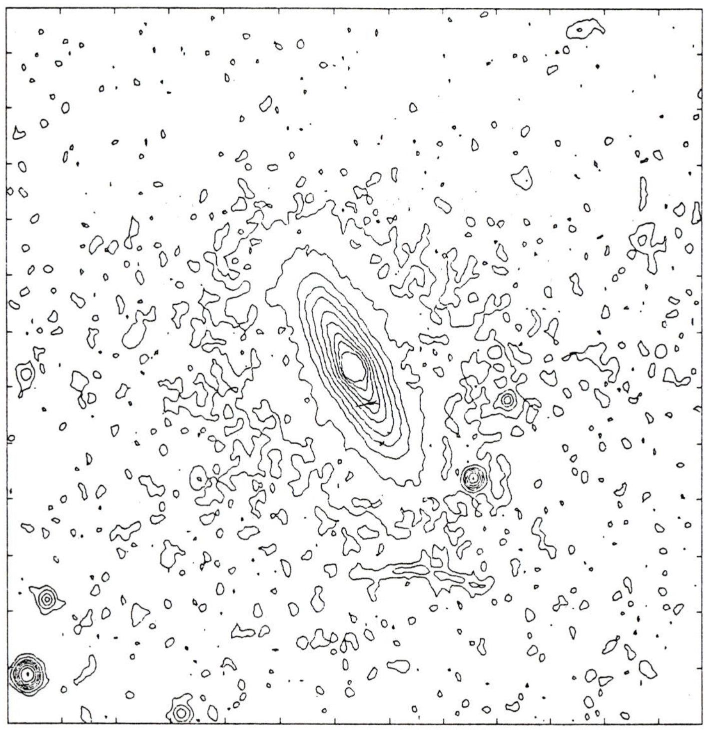

A2.9b: NGC 4350 SA0sp (I)

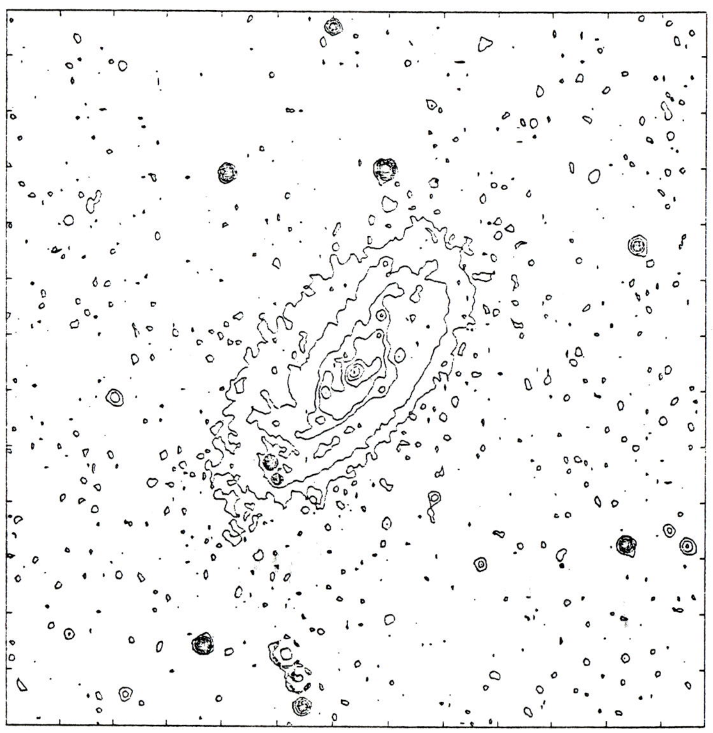

A2.10b: NGC 4501 SA(rs)b (U)

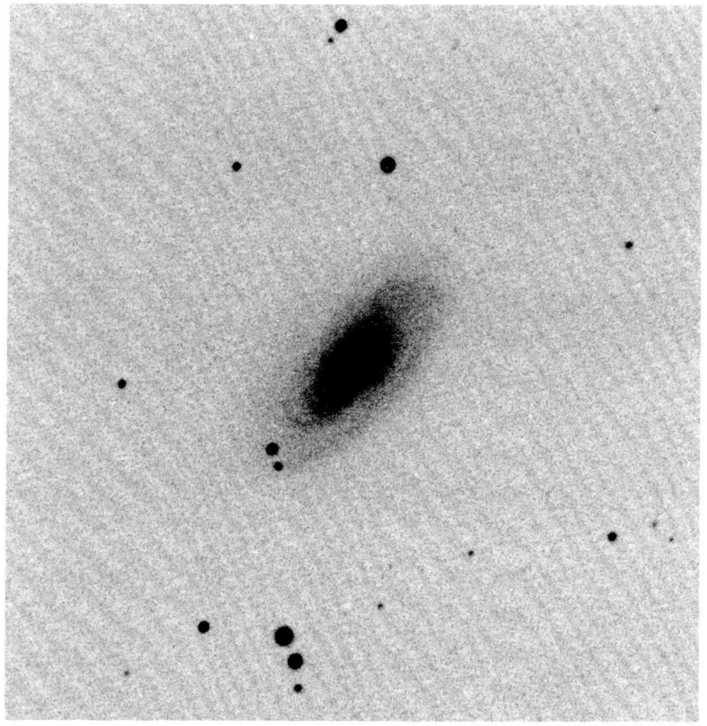

A2.11a: NGC 4501 SA(rs)b (I)

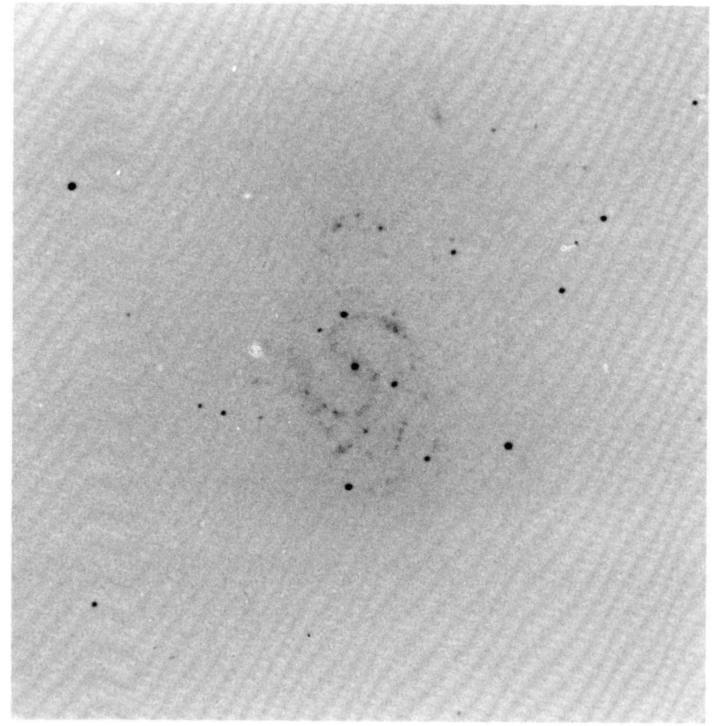

A2.12a: NGC 4535 SAB(s)c (U)

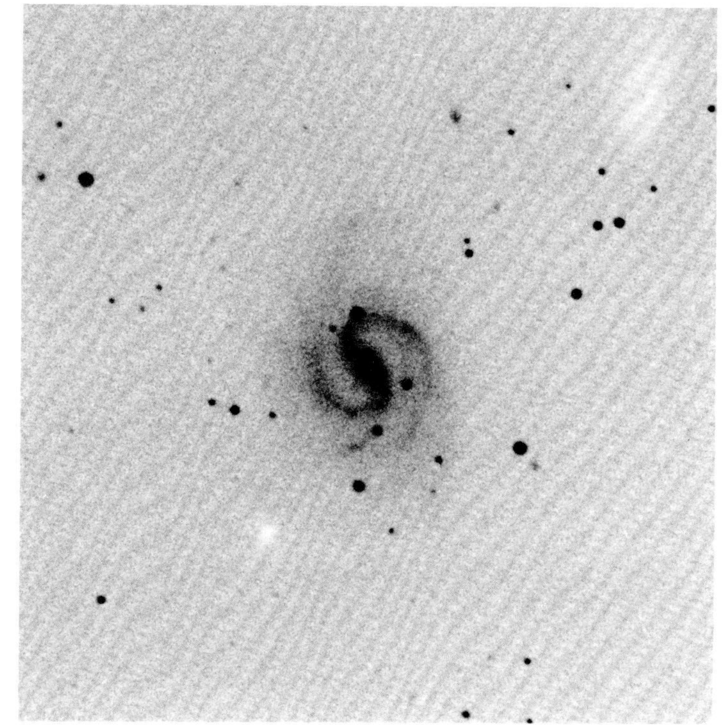

A2.13a: NGC 4535 SAB(s)c (I)

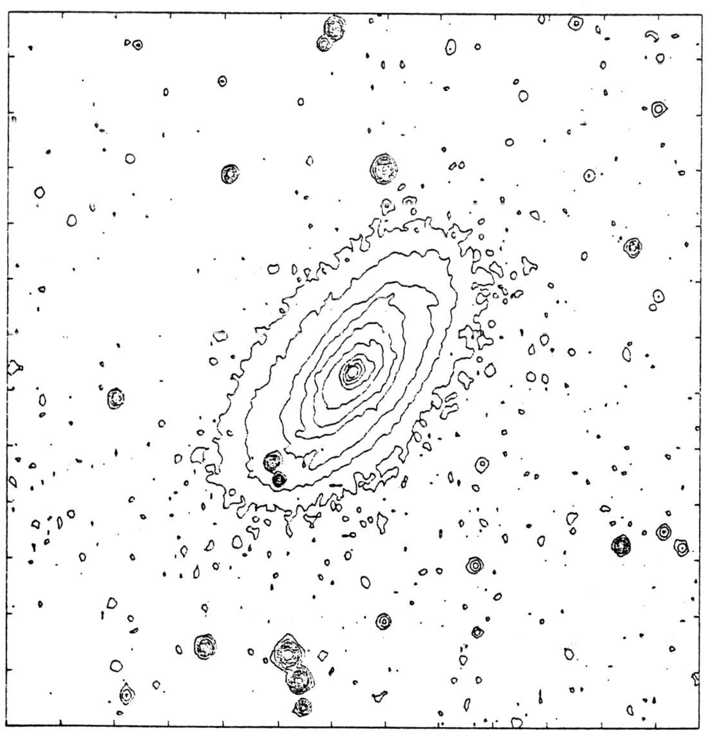

A2.11b: NGC 4501 SA(rs)b (I)

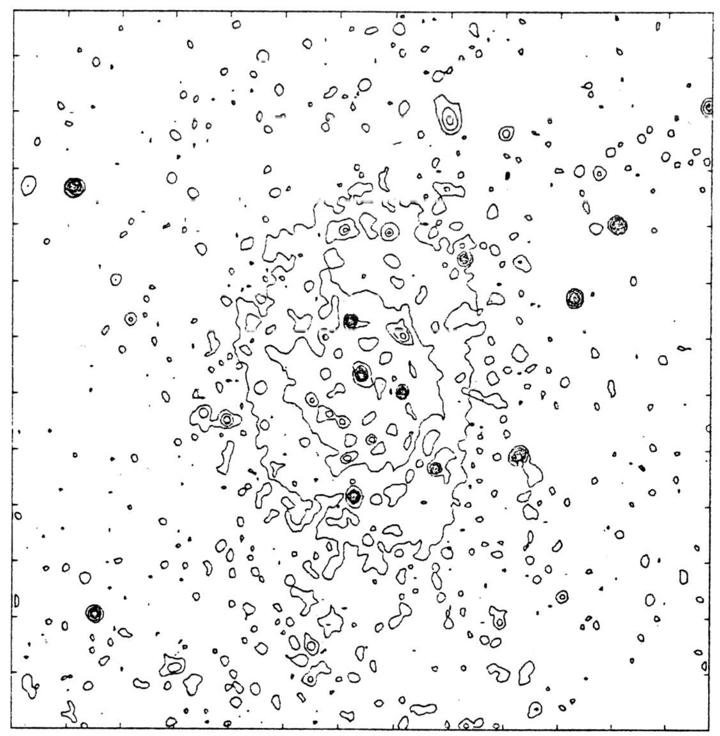

A2.12b: NGC 4535 SAB(s)c (U)

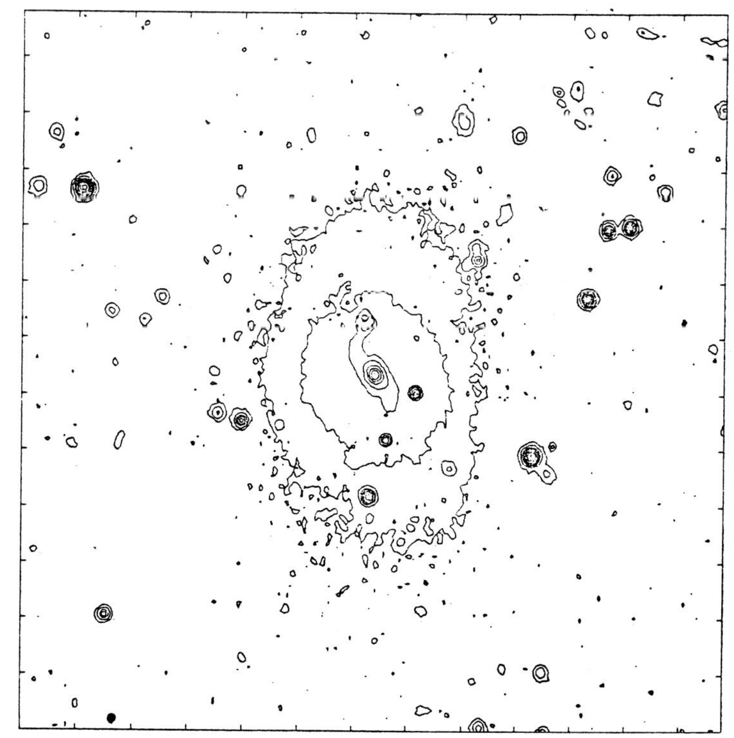

A2.13b: NGC 4535 SAB(s)c (I)

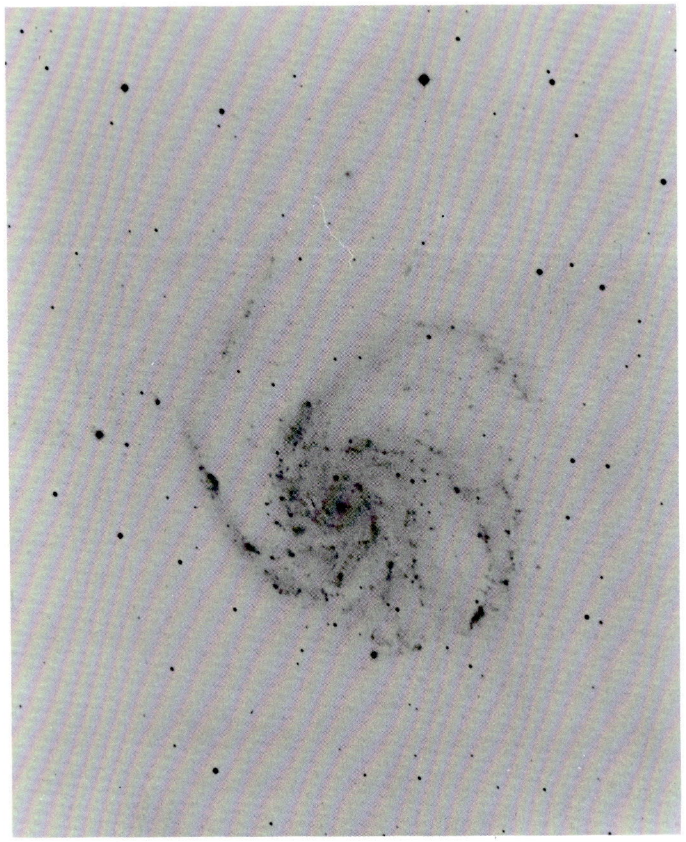

A2.14a: NGC 5457 (M101) SAB(rs)cd (U)

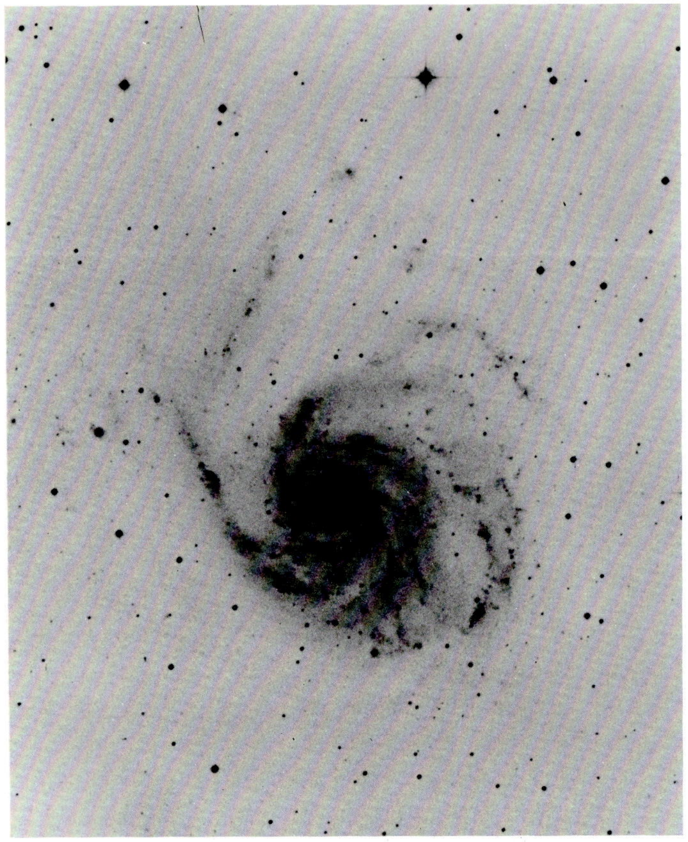

A2.14b: NGC 5457 (M101) SAB(rs)cd (B)

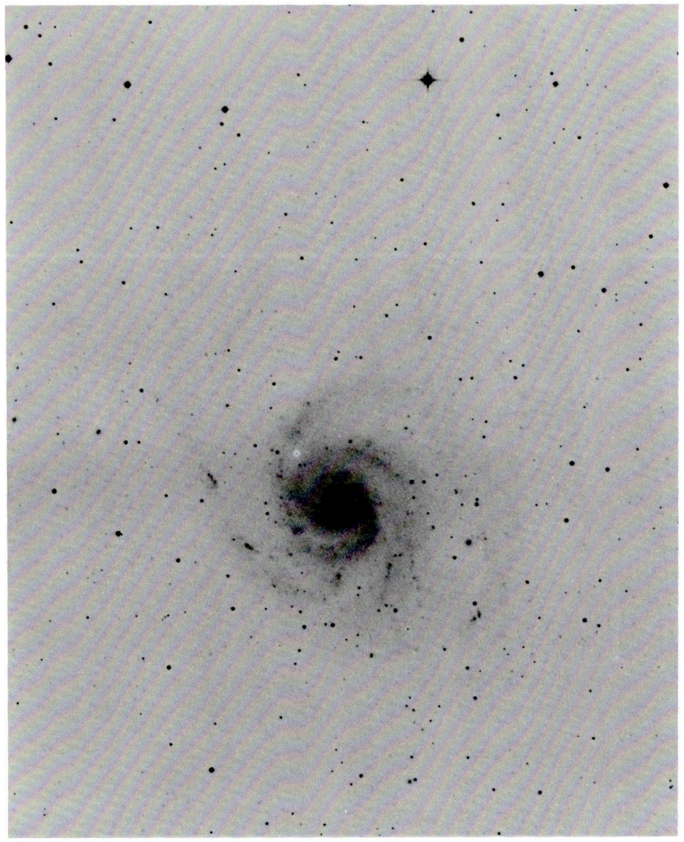

A2.14c: NGC 5457 (M101) SAB(rs)cd (I)

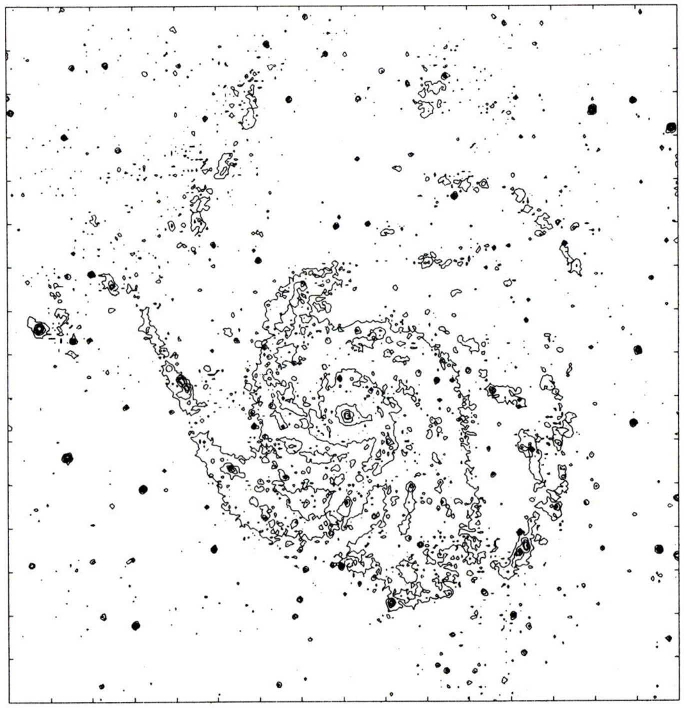

A2.14d: NGC 5457 (M101) SAB(rs)cd (B)

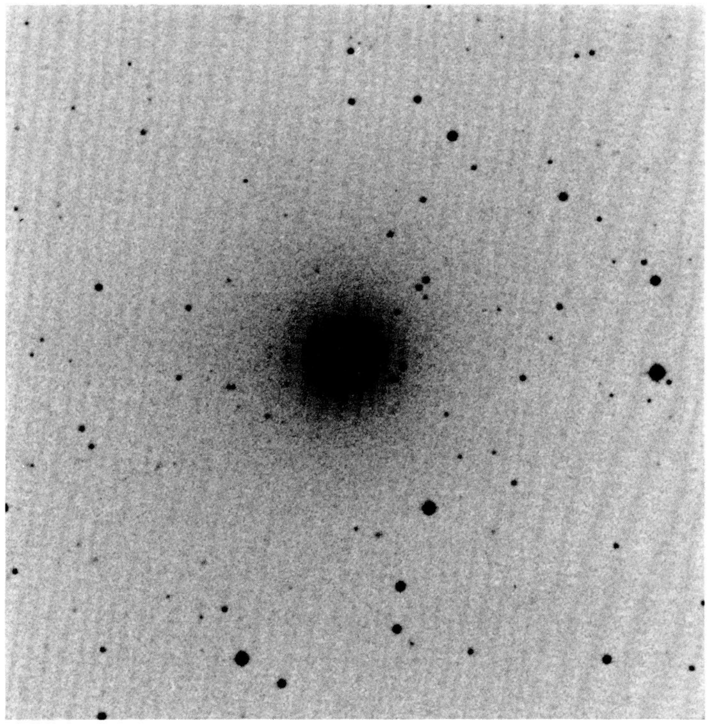

A3.1a: NGC 4374 E1 (B)

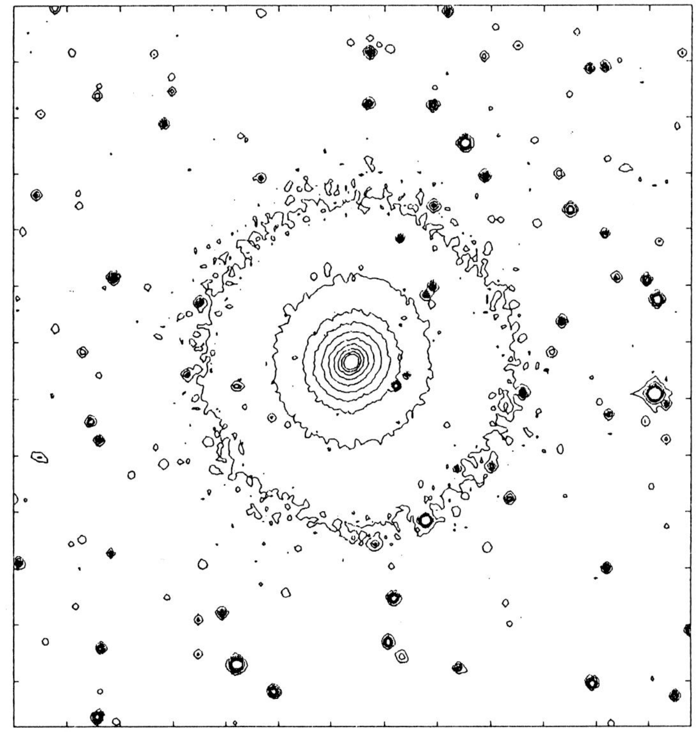

A3.1b: NGC 4374 E1 (V)

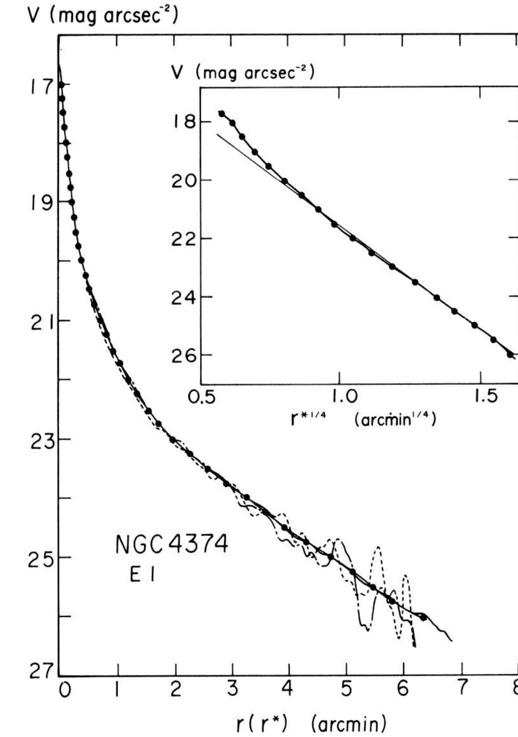


A3.1c: Luminosity profiles

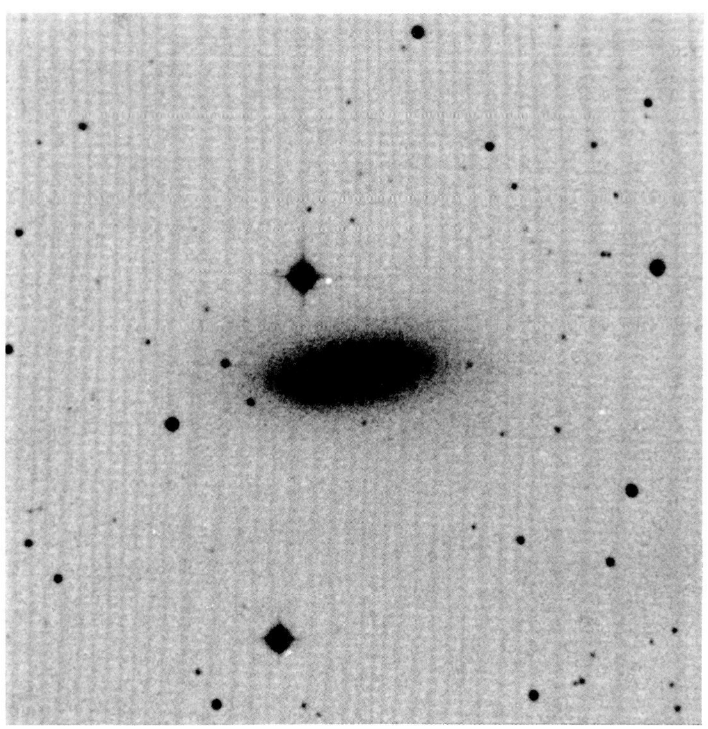

A3.2a: NGC 4429 SA(r)0$^+$ (B)

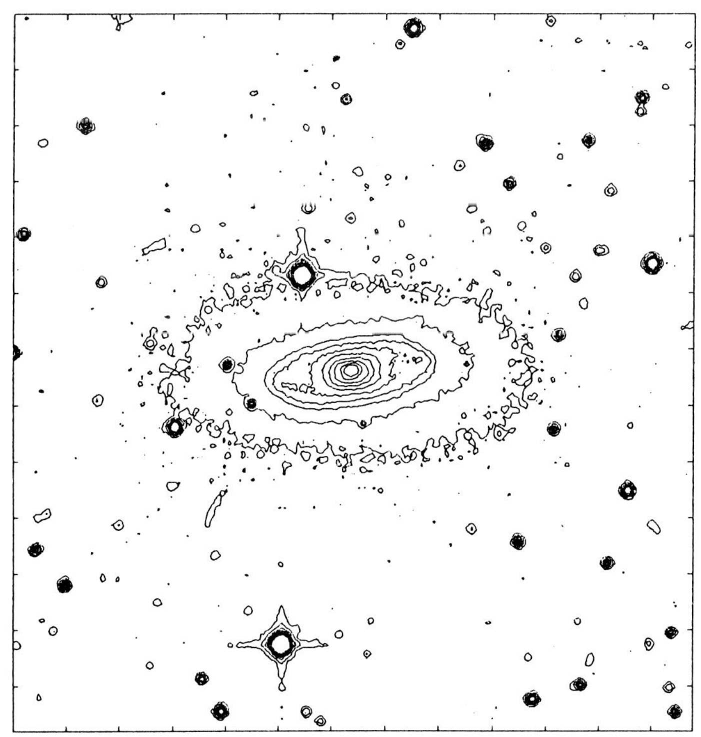

A3.2b: NGC 4429 SA(r)0$^+$ (V)

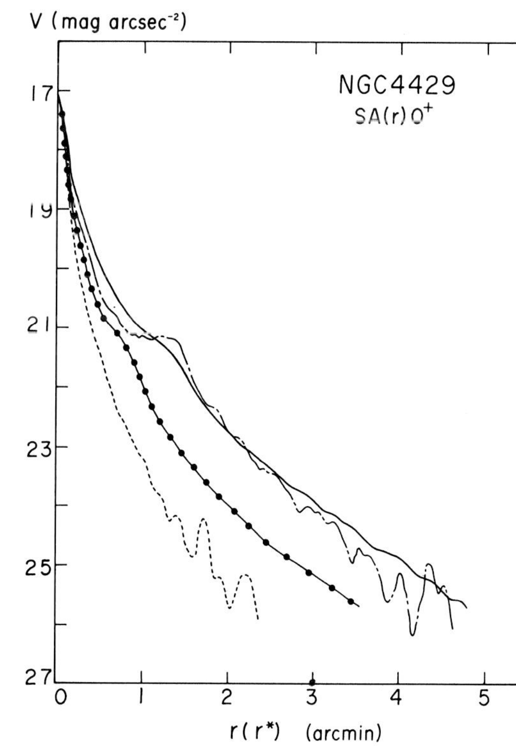


A3.2c: Luminosity profiles

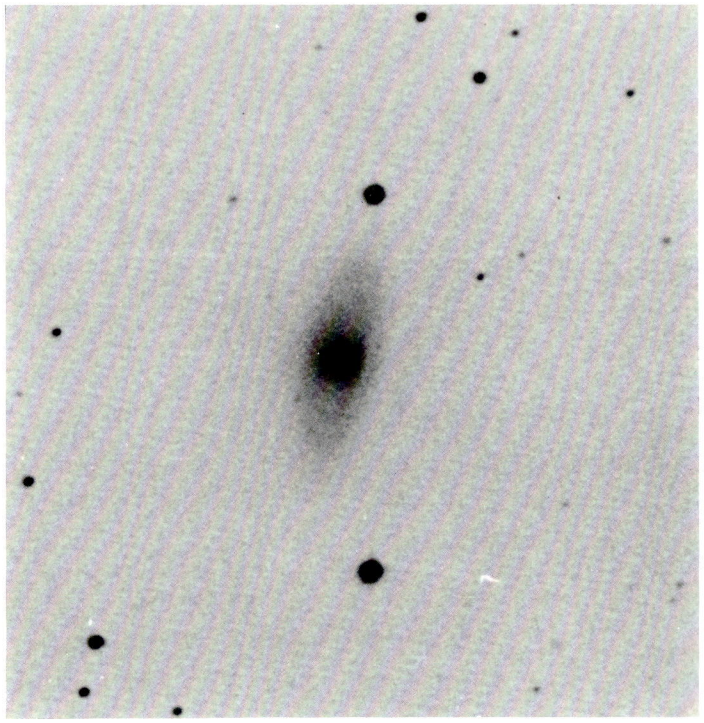

A3.3a: NGC 4698 SA(s)ab (B)

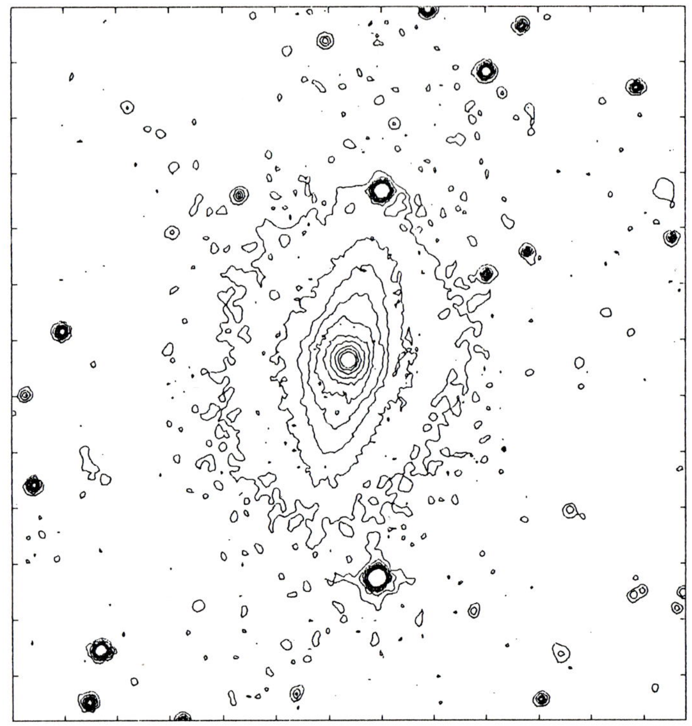

A3.3b: NGC 4698 SA(s)ab (V)

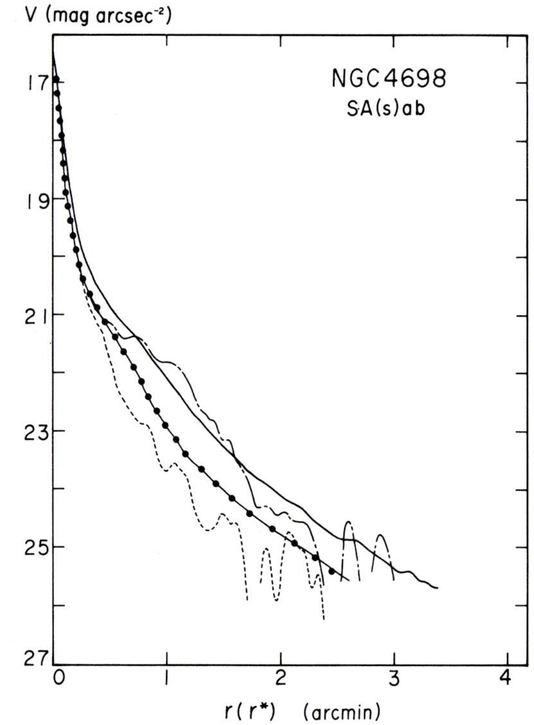


A3.3c: Luminosity profiles

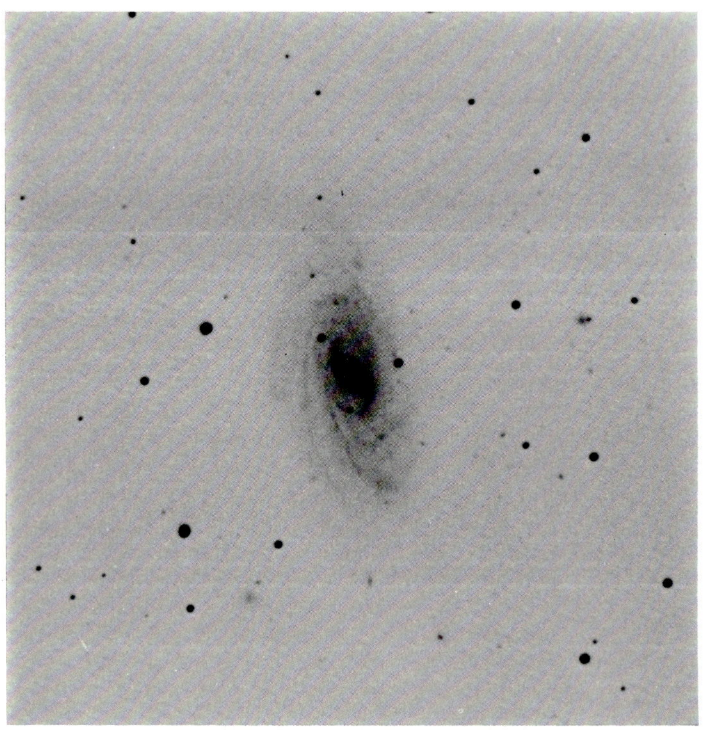

A3.4a: NGC 3953 SB(r)bc (V)

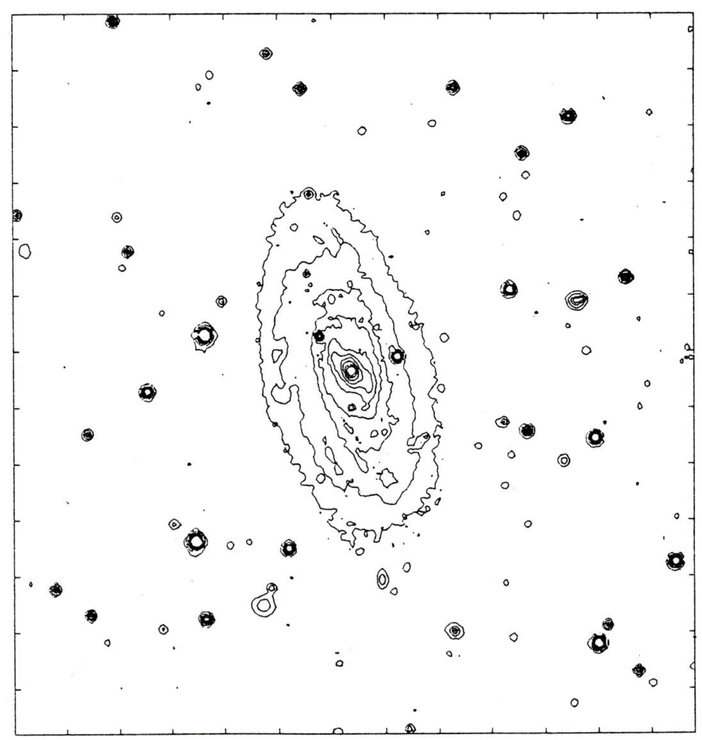

A3.4b: NGC 3953 SB(r)bc (V)

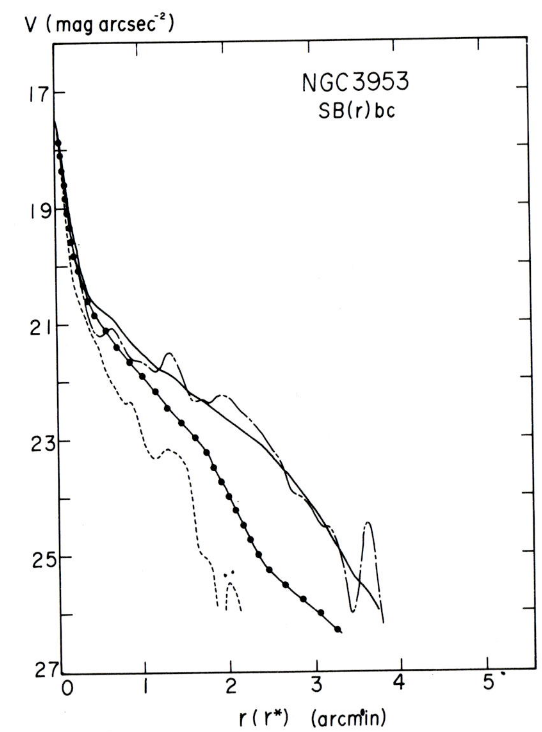


A3.4c: Luminosity profiles

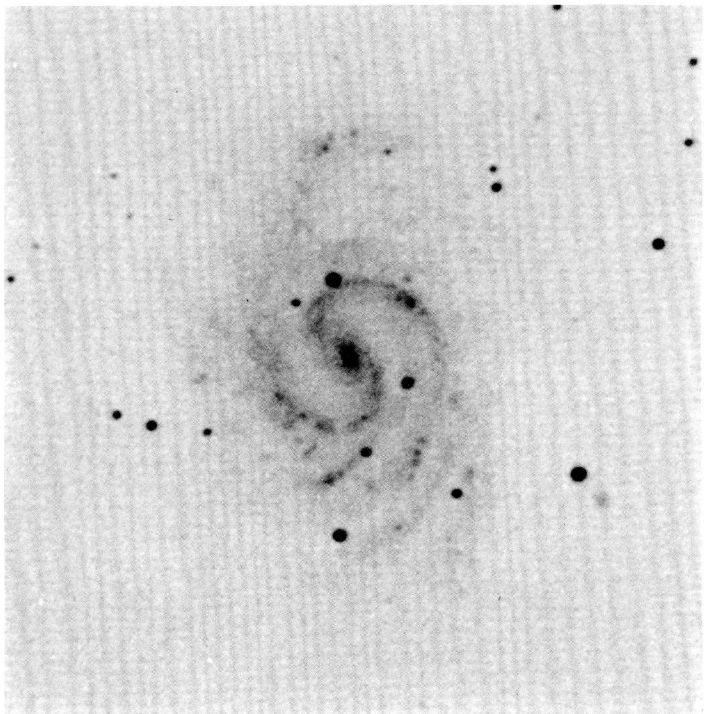

A3.5a: NGC 4535 SAB(s)c (B)

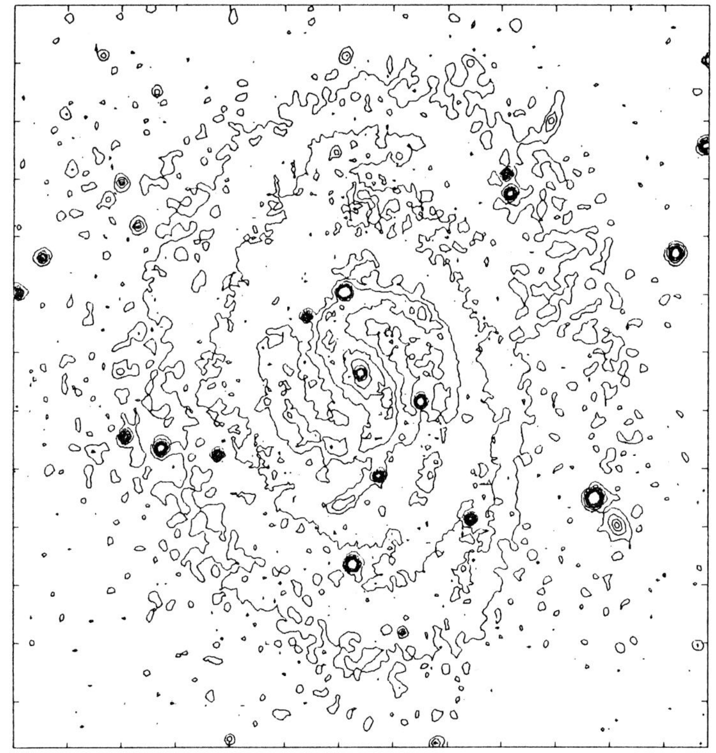

A3.5b: NGC 4535 SAB(s)cd (V)

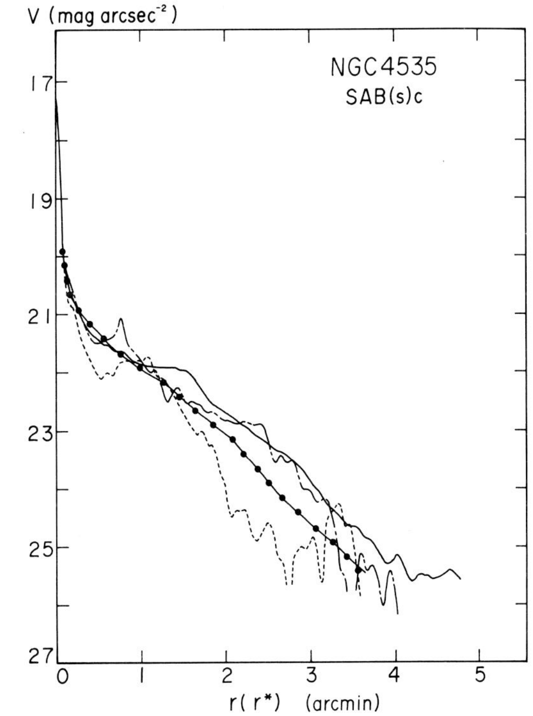


A3.5c: Luminosity profiles

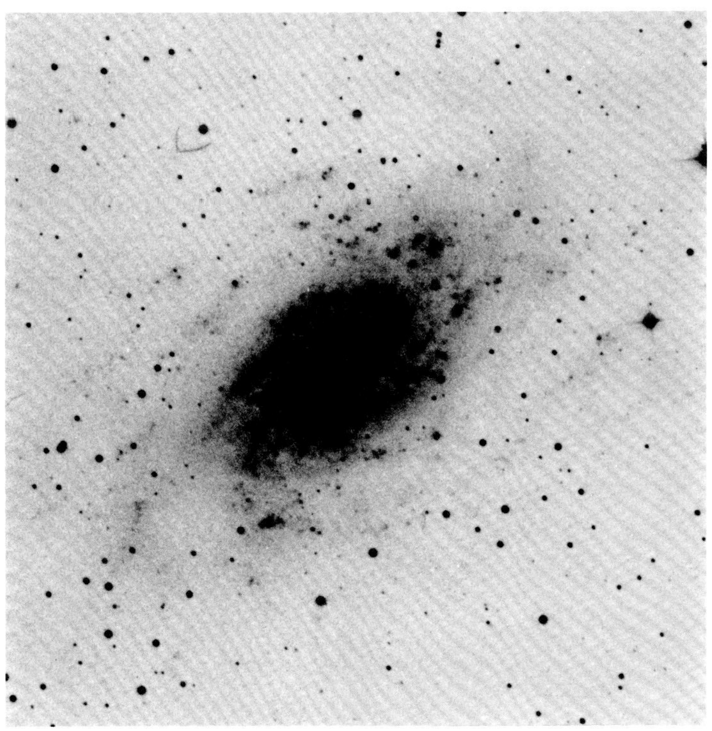

A3.6a: NGC 2403 SAB(s)cd (B)

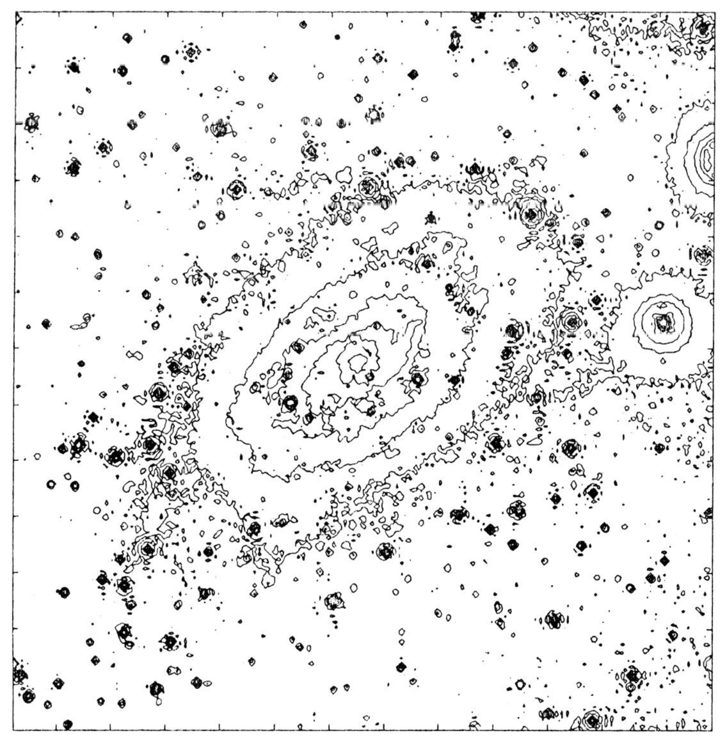

A3.6b: NGC 2403 SAB(s)cd (V)

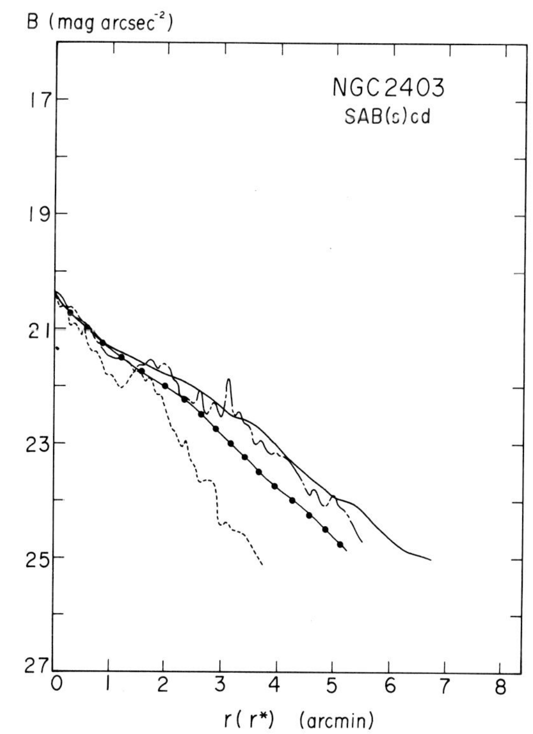


A3.6c: Luminosity profiles

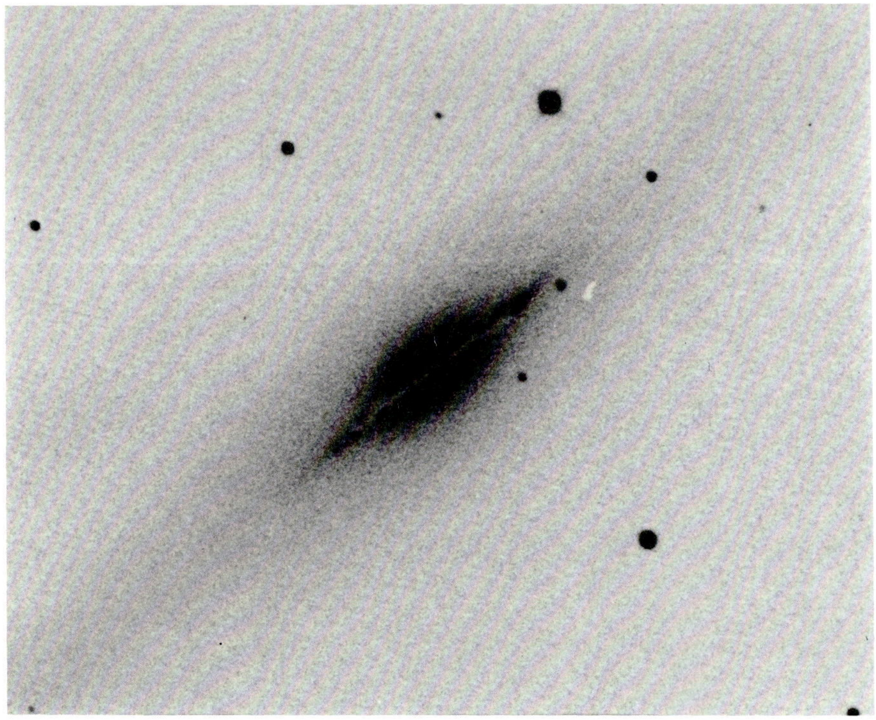

A4.1a: NGC 5866 SA0⁺sp (B)

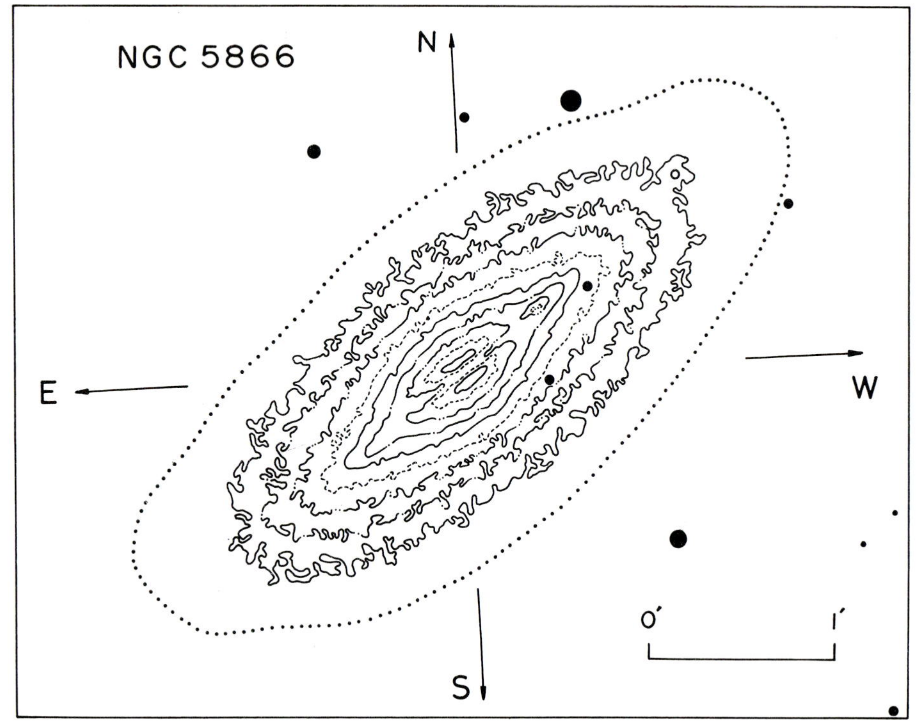


A4.1b: NGC 5866 SA0⁺sp (B)

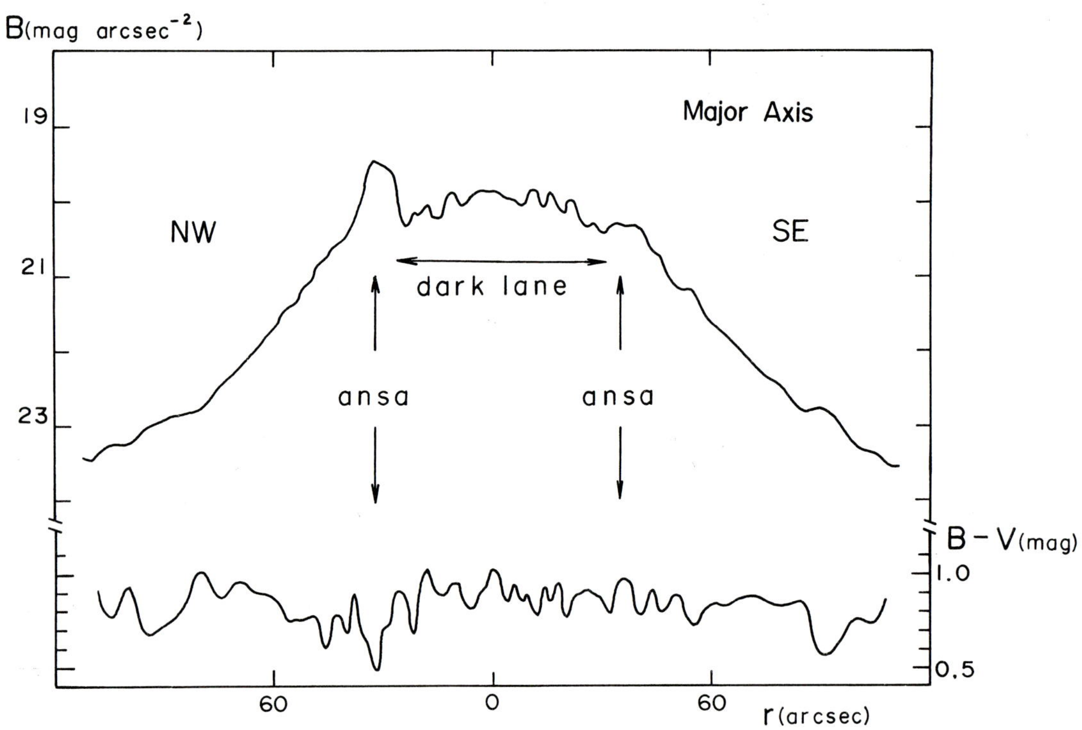


A4.1c: Major-axis profile

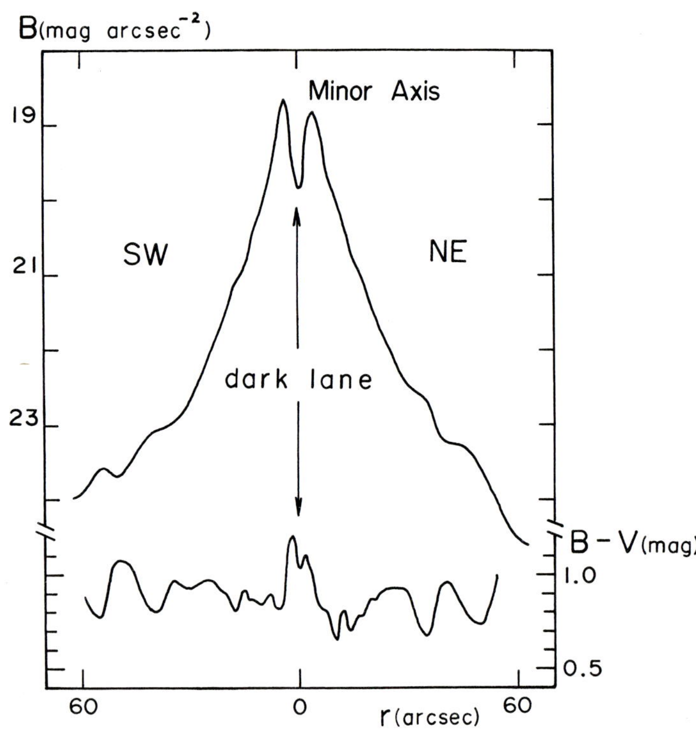


A4.1d: Minor-axis profile

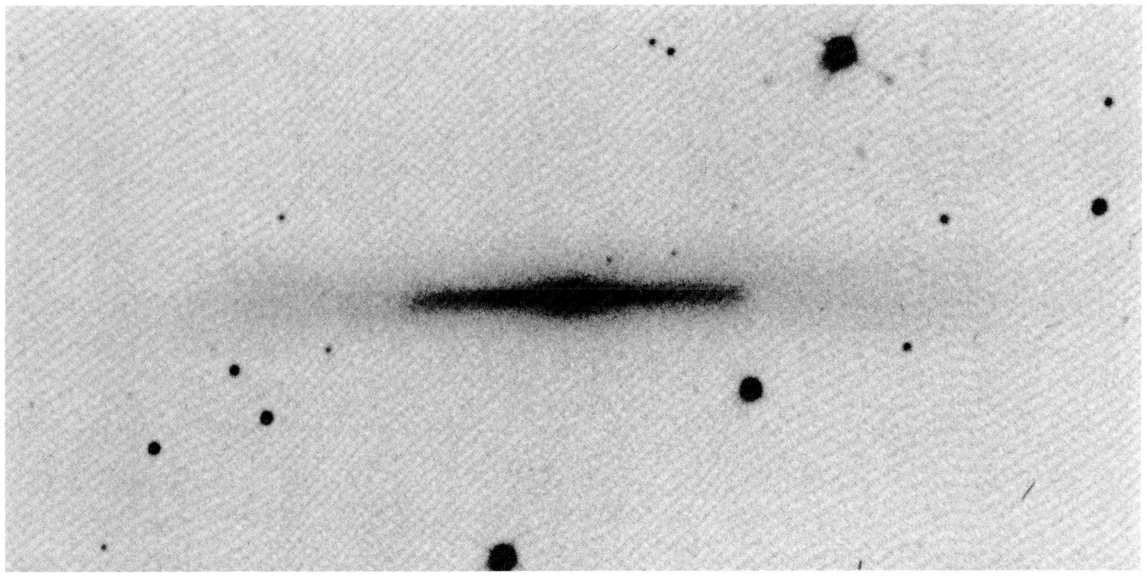

A4.2a: NGC 4762 SB(r)0$^0$?sp (B)

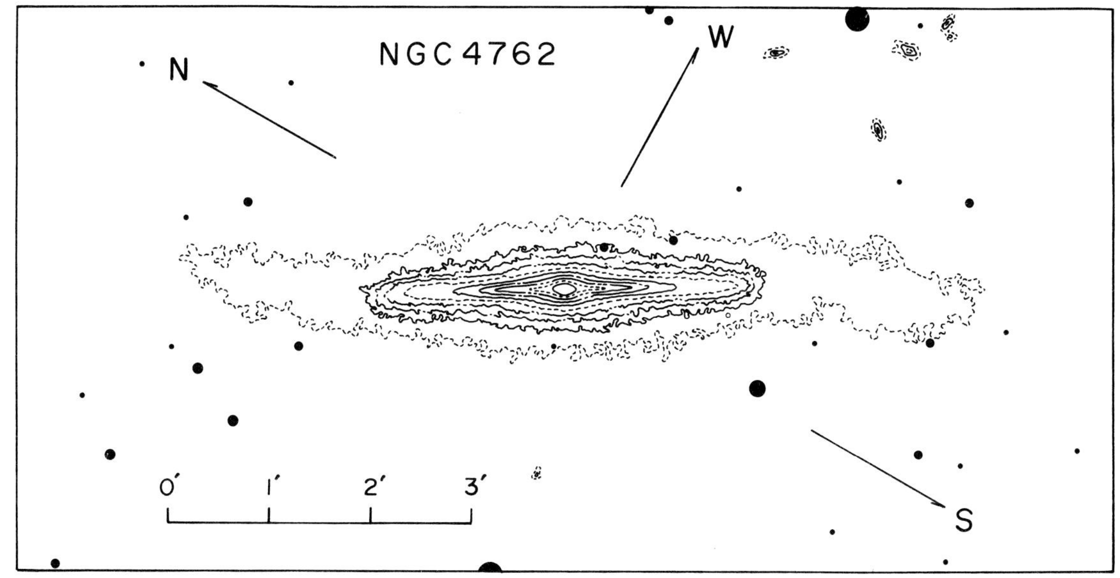


A4.2b: NGC 4762 SB(r)0$^0$?sp (B)

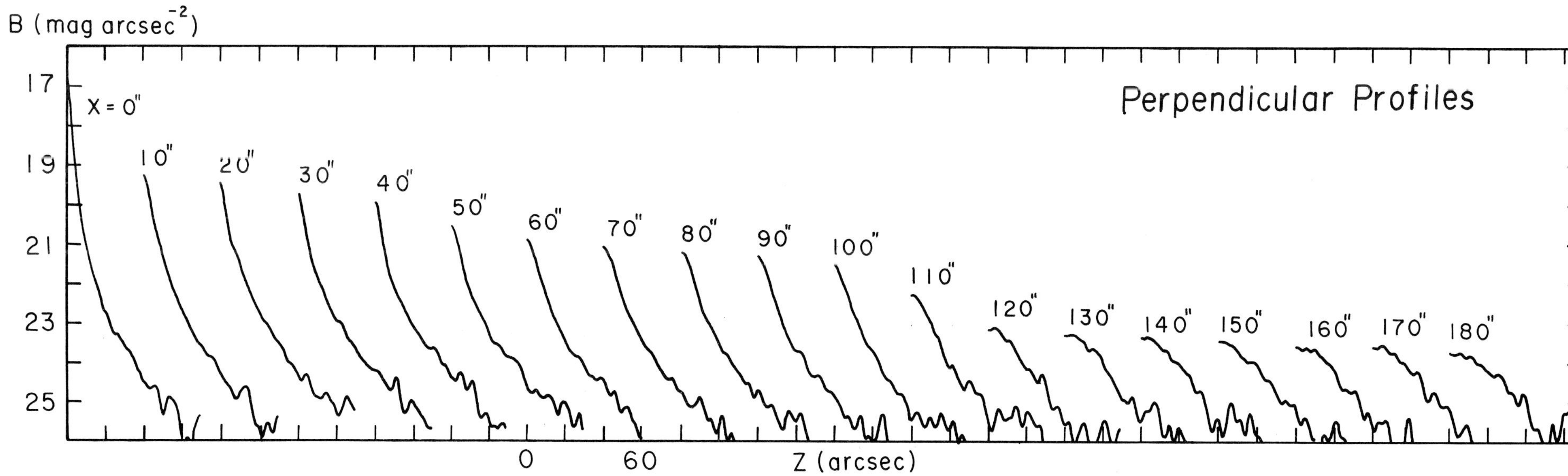


A4.2c: Profiles perpendicular to the major axis at various distances from the galactic center

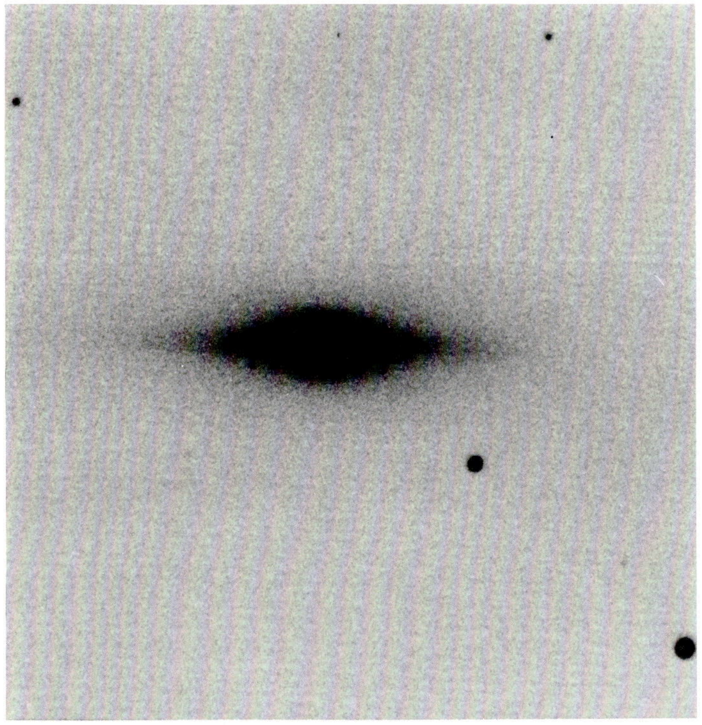

A4.3a: NGC 3115 S0$^-$sp (V)

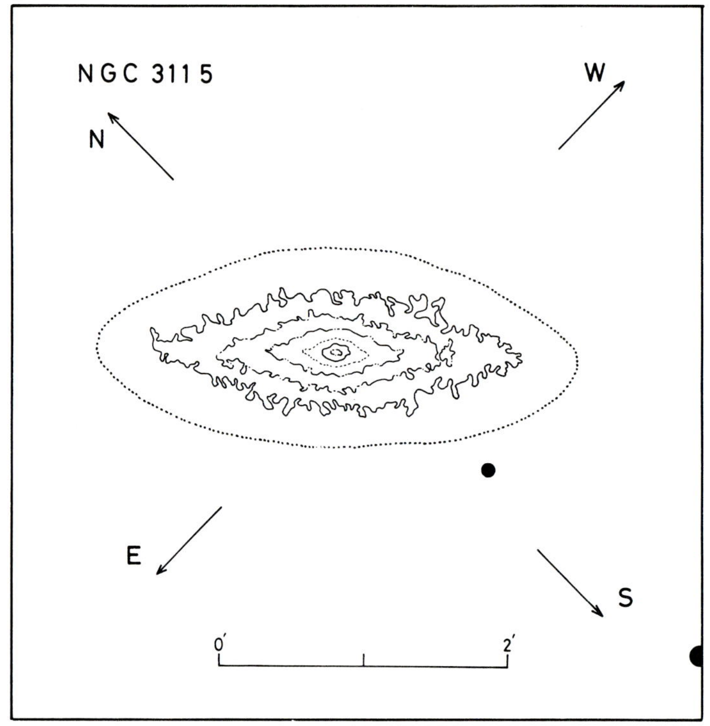


A4.3b: NGC 3115 S0$^-$sp (B)

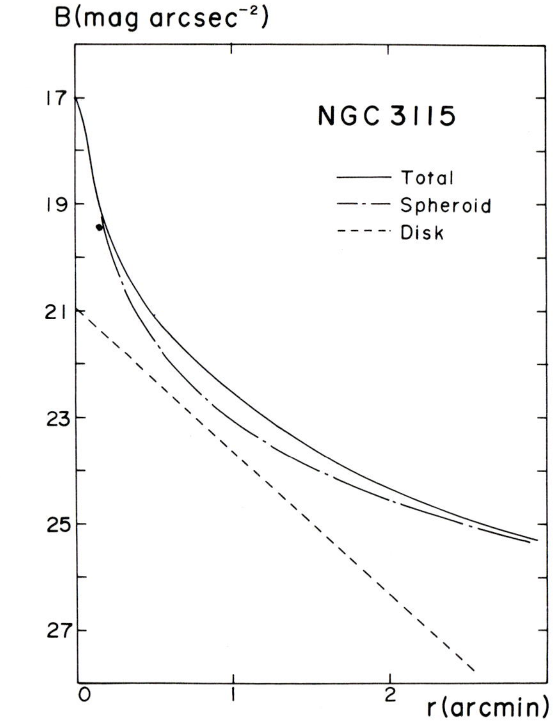


A4.3c: Face-on profiles

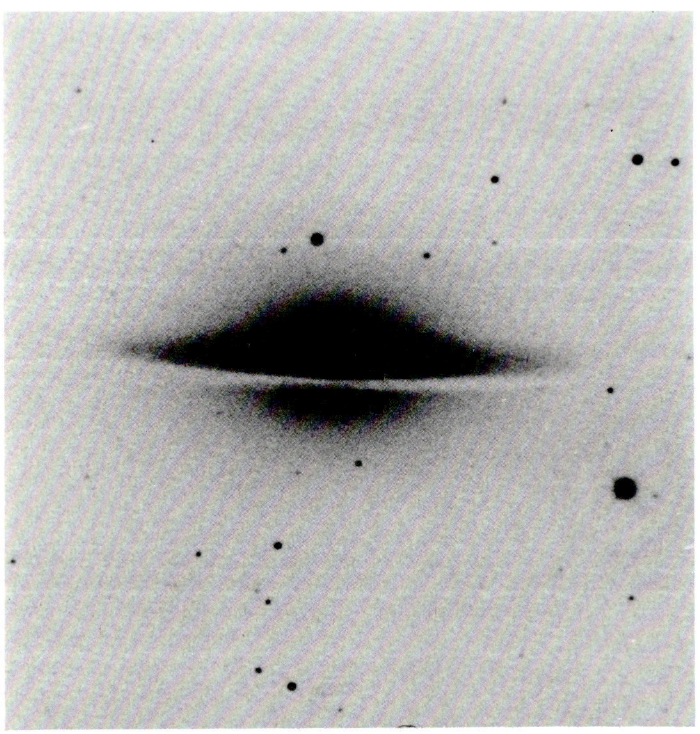

A4.4a: NGC 4594 SA(s)a sp (V)

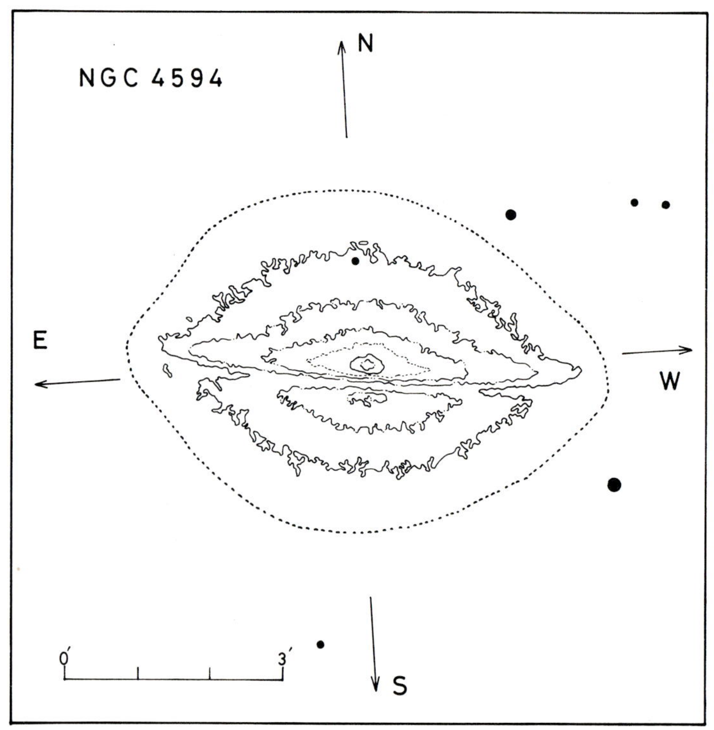


A4.4b: NGC 4594 SA(s)a sp (B)

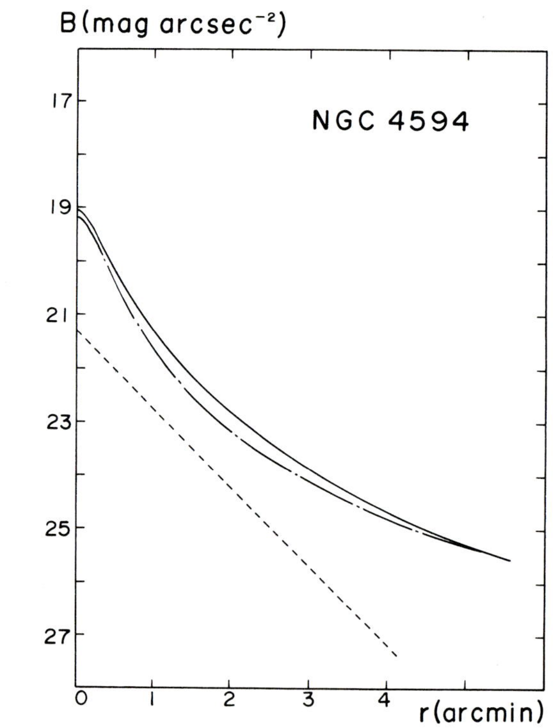


A4.4c: Face-on profiles

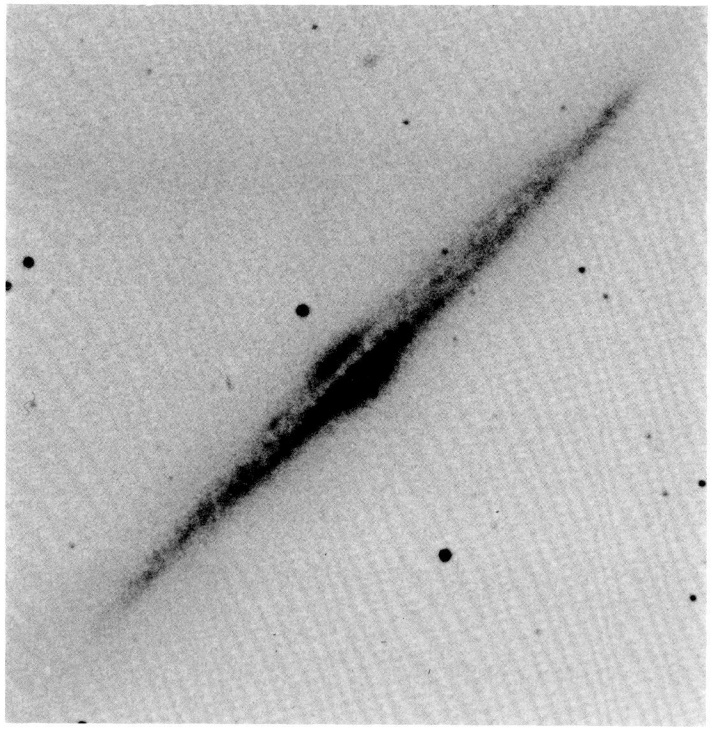

A4.5a: NGC 4565 SA(s)b?sp (B)

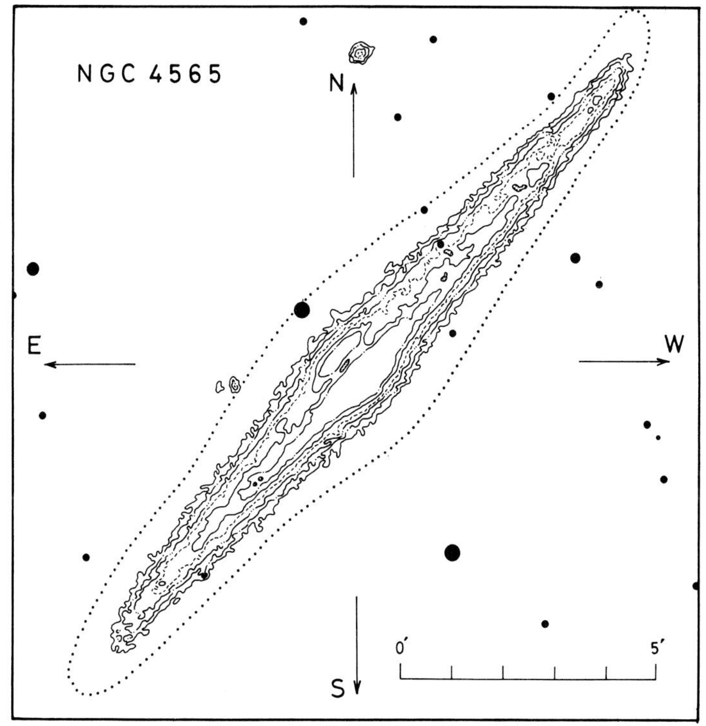


A4.5b: NGC 4565 SA(s)b?sp (B)

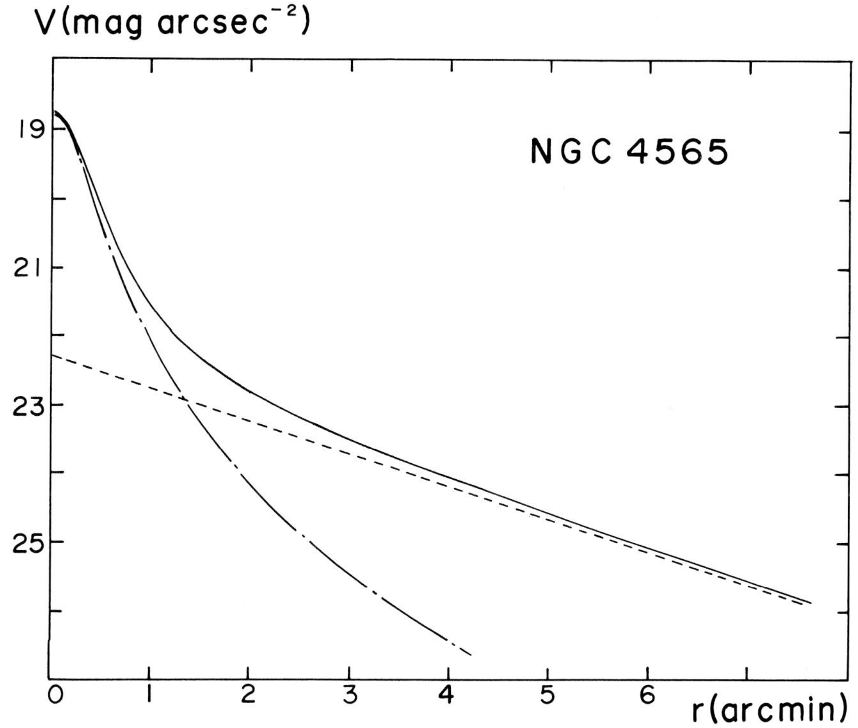


A4.5c: Face-on profiles

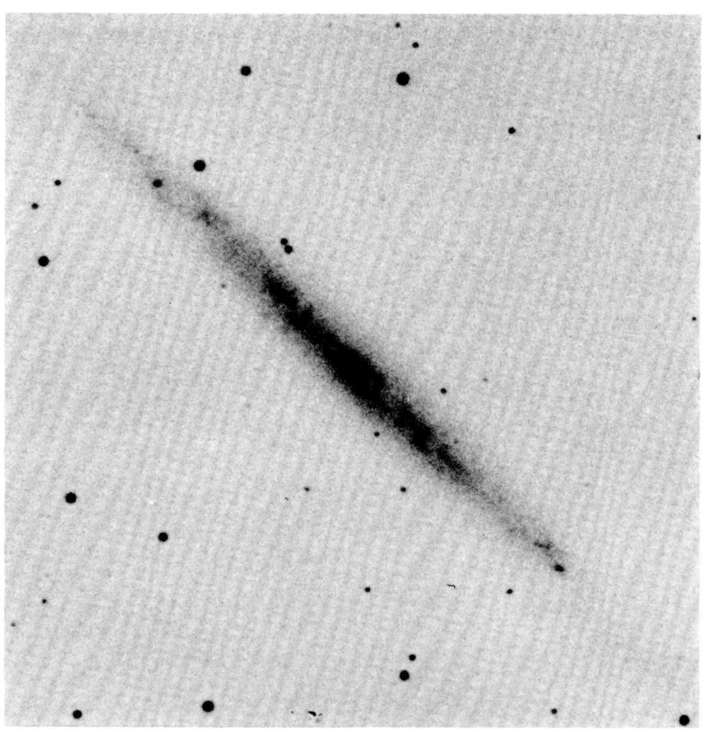

A4.6a: NGC 4244 SA(s)cd:sp (B)

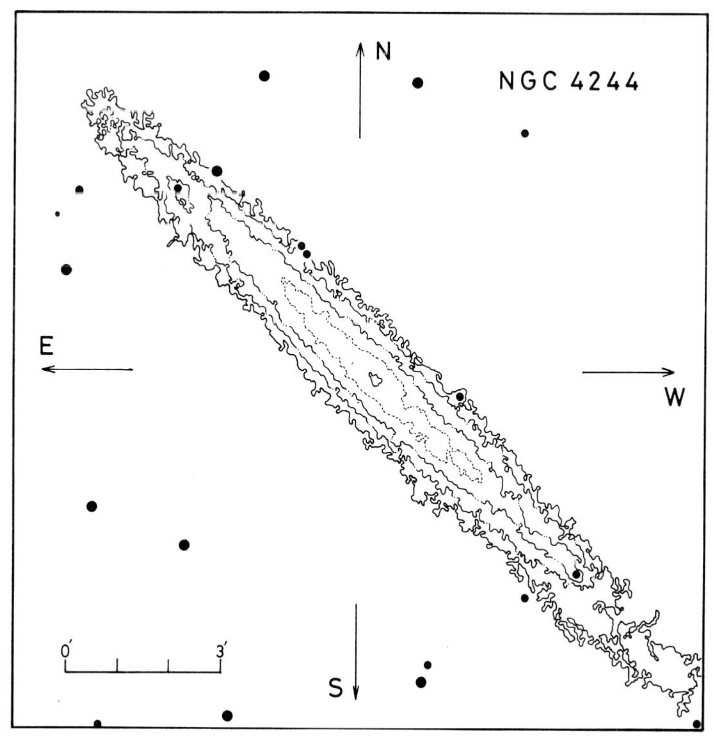


A4.6b: NGC 4244 SA(s)cd:sp (B)

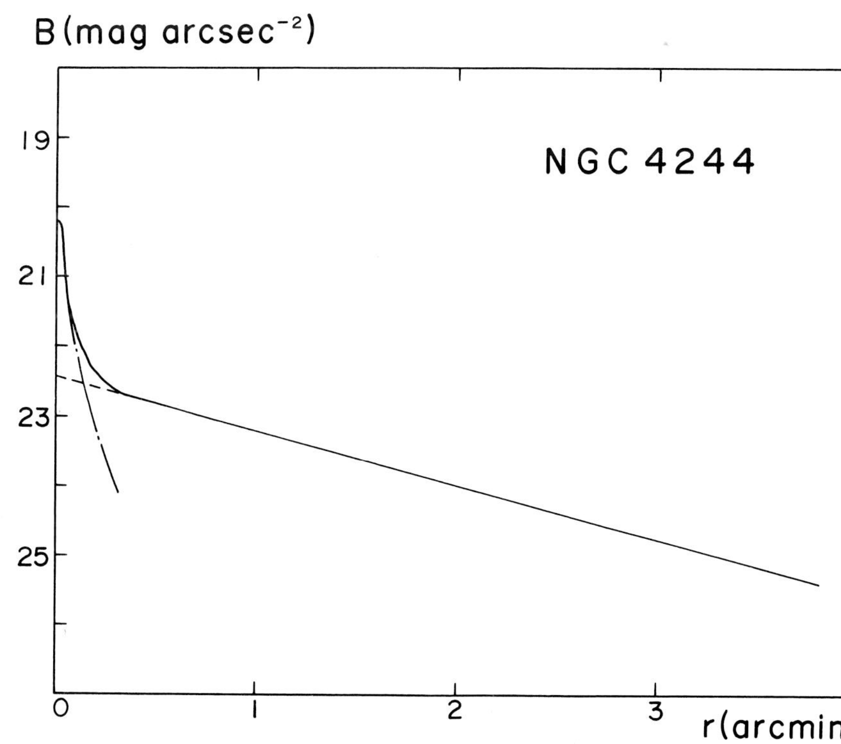


A4.6c: Face-on profiles

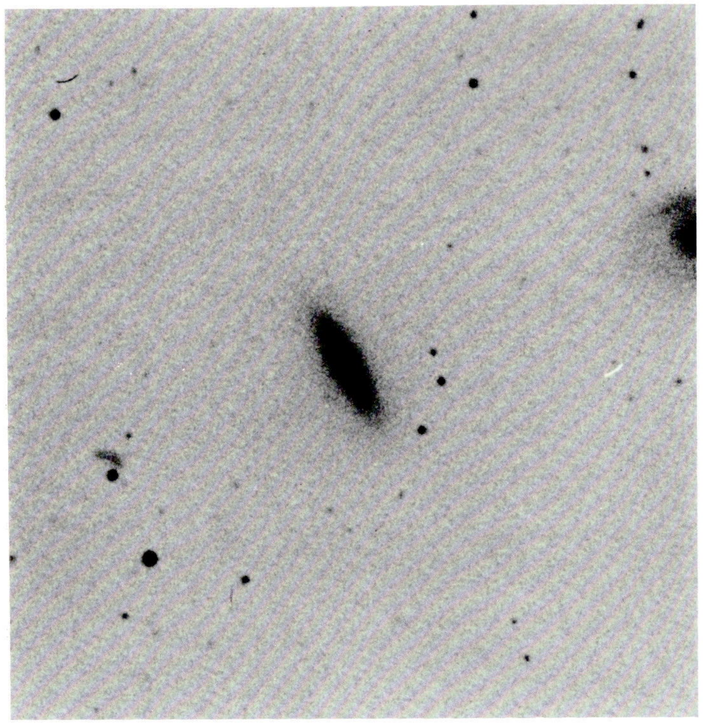

A4.7: NGC 4350 SA0sp (B)

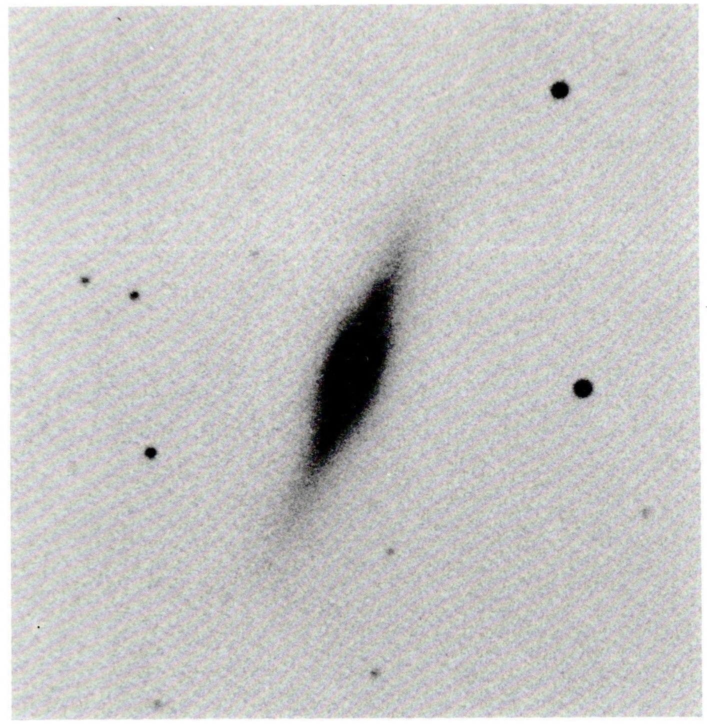

A4.8: NGC 7332 S0Psp (V)

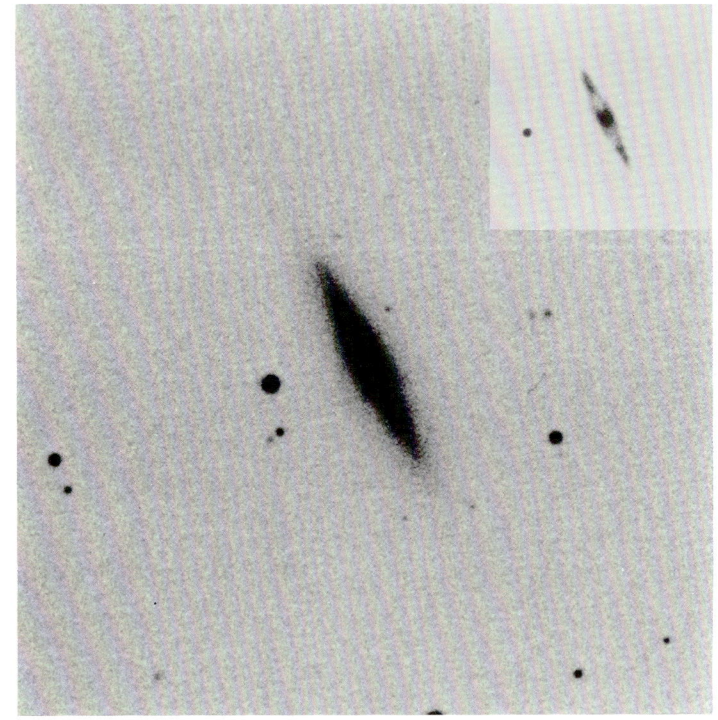

A4.9: NGC 4710 SA(r)$0^+$?sp (B)

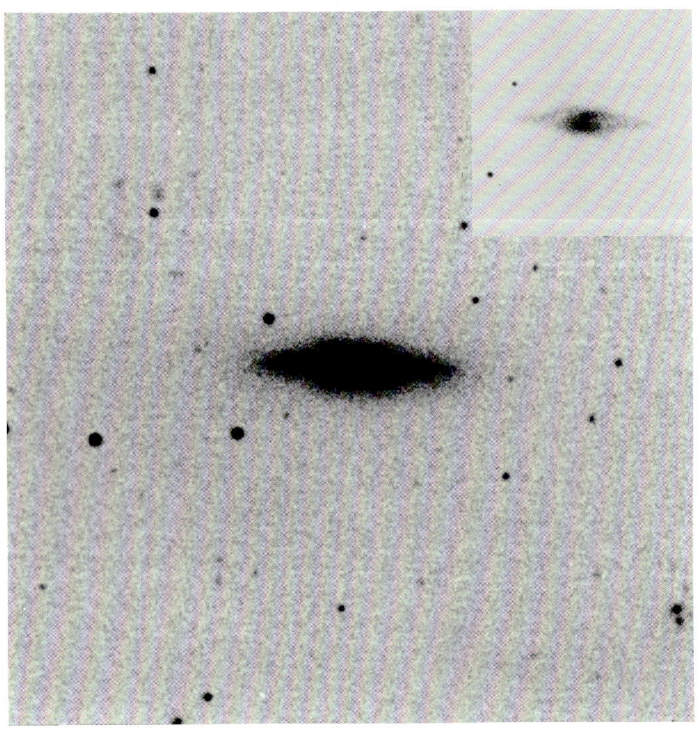

A4.10: NGC 4469 SB(s)0/a?sp (B)

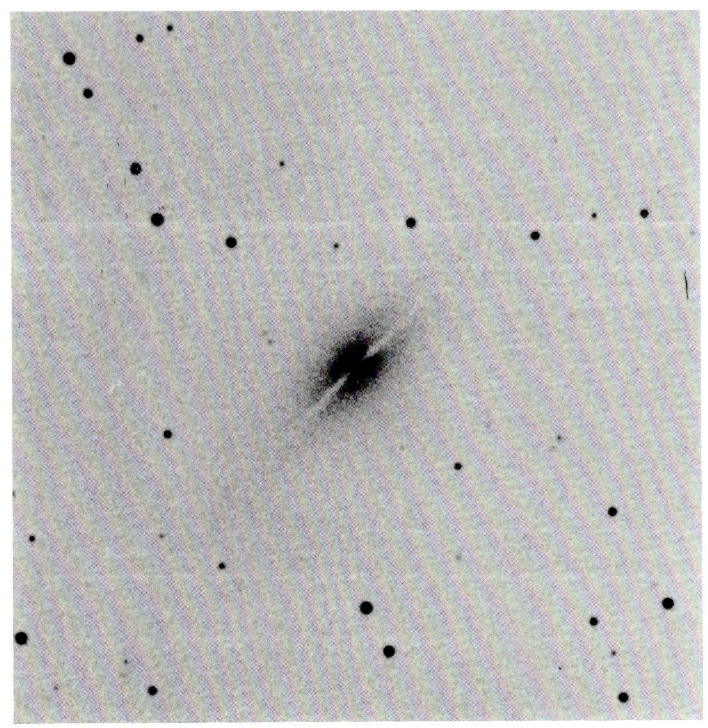

A4.11: NGC 7814 SA(s)ab:sp (B)

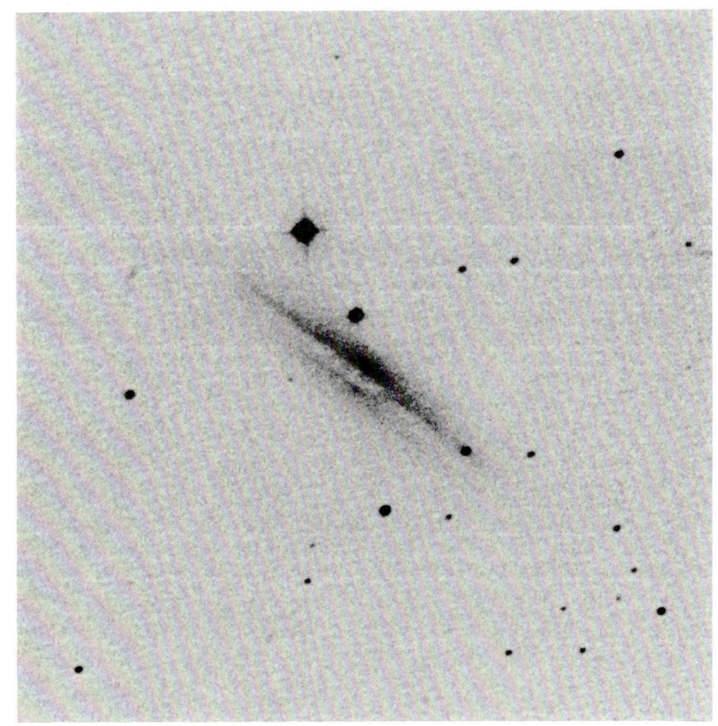

A4.12: NGC 4217 Sb sp (B)

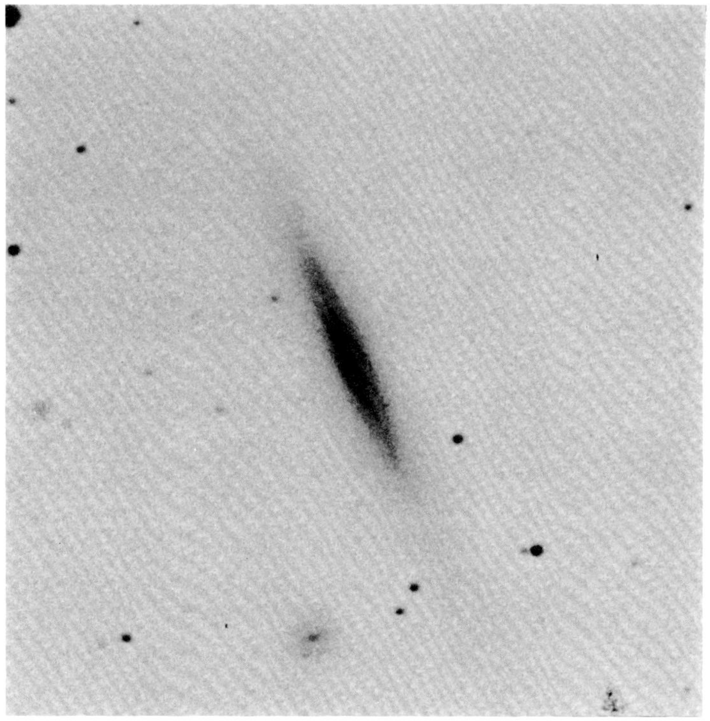

A4.13: NGC 4307 Sb sp (V)

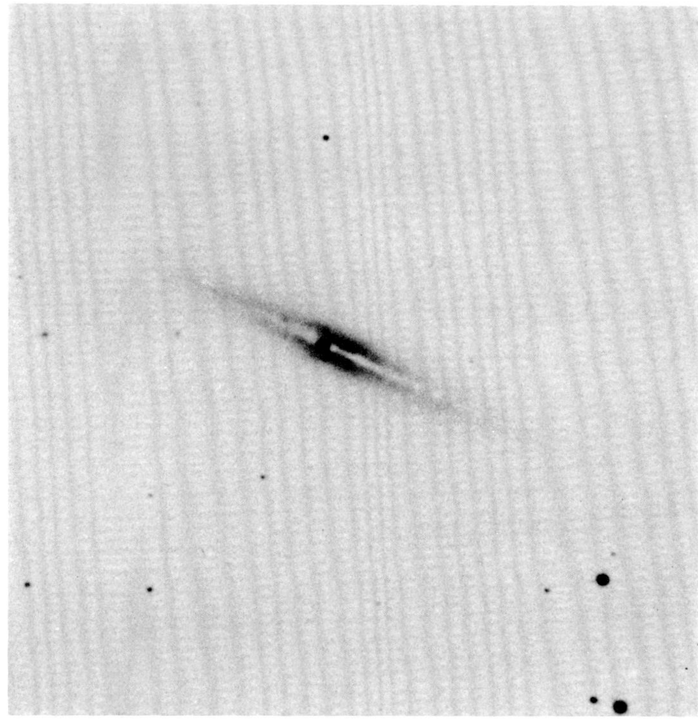

A4.14: NGC 4013 Sb sp (V)

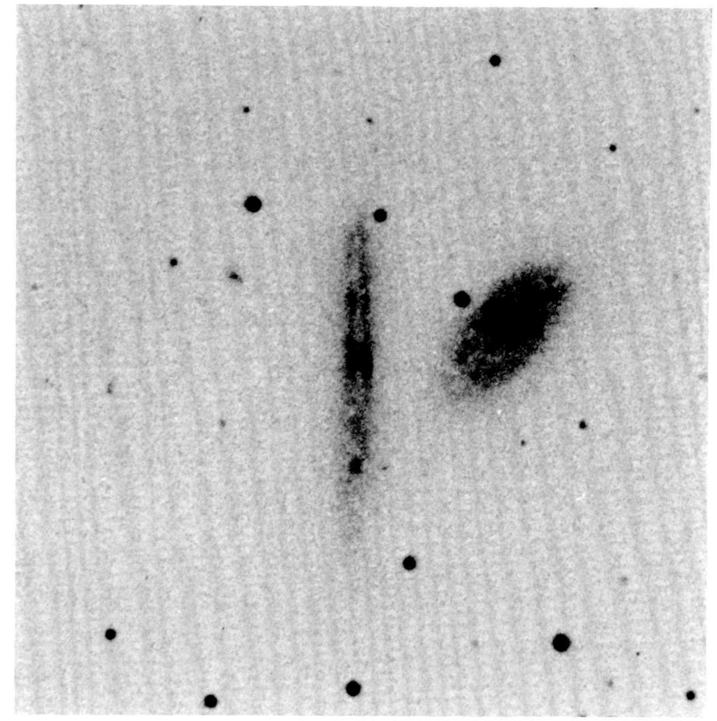

A4.15: NGC 4302 Sc:sp,
NGC 4298 SA(rs)c (B)

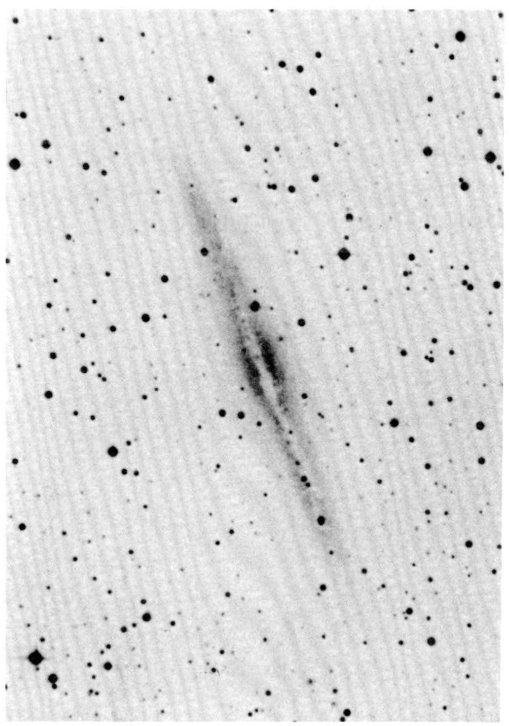

A4.16: NGC 891 SA(s)b?sp (B)

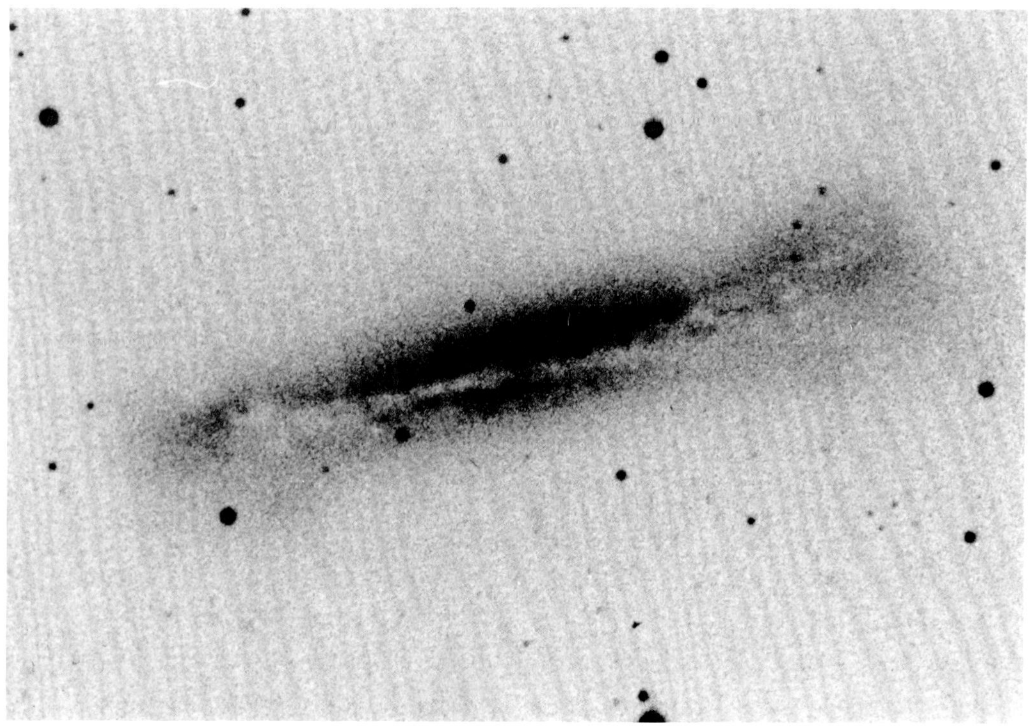

A4.17: NGC 3628 SbP sp (B)

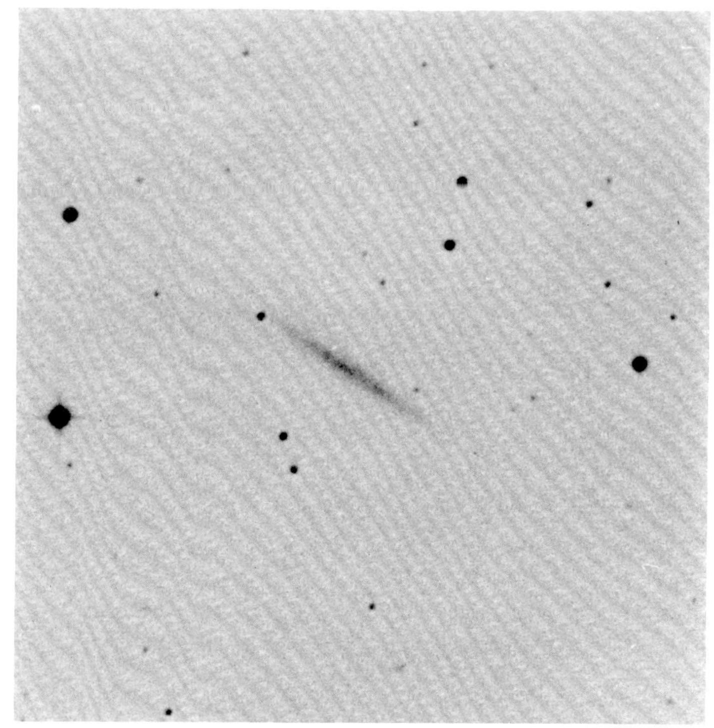

A4.18: NGC 4222 Sd:sp (B)

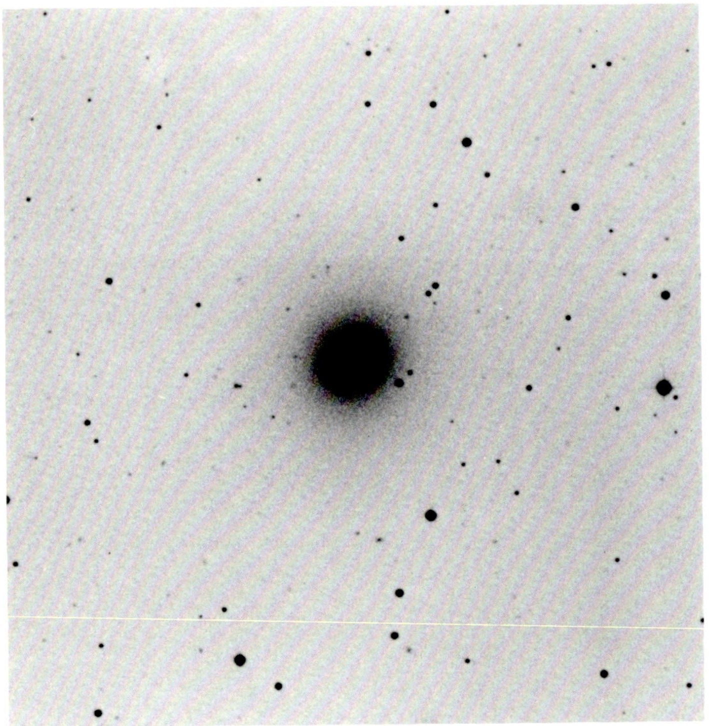

A5.1a: NGC 4374 E1 (V) $X1(P)$=1.33

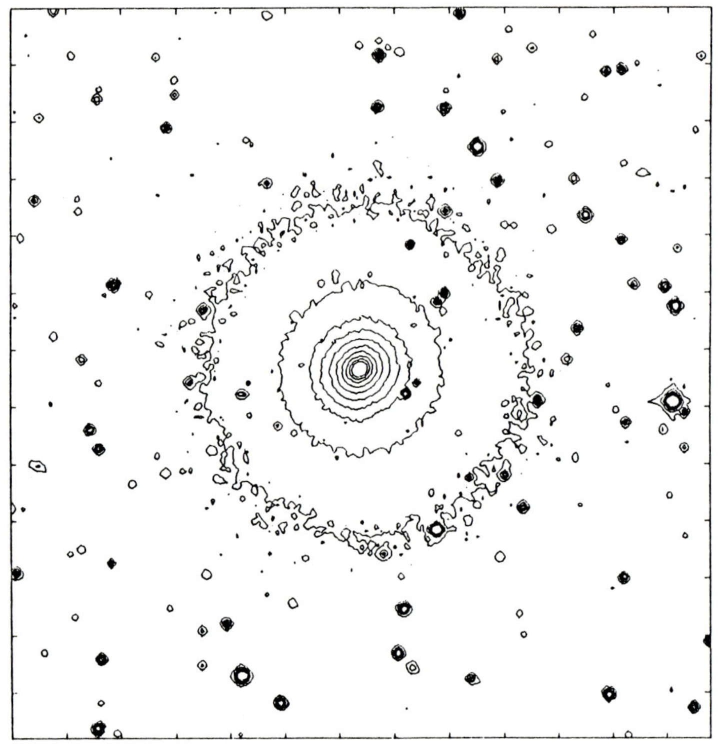

A5.1b: NGC 4374 E1 (V)

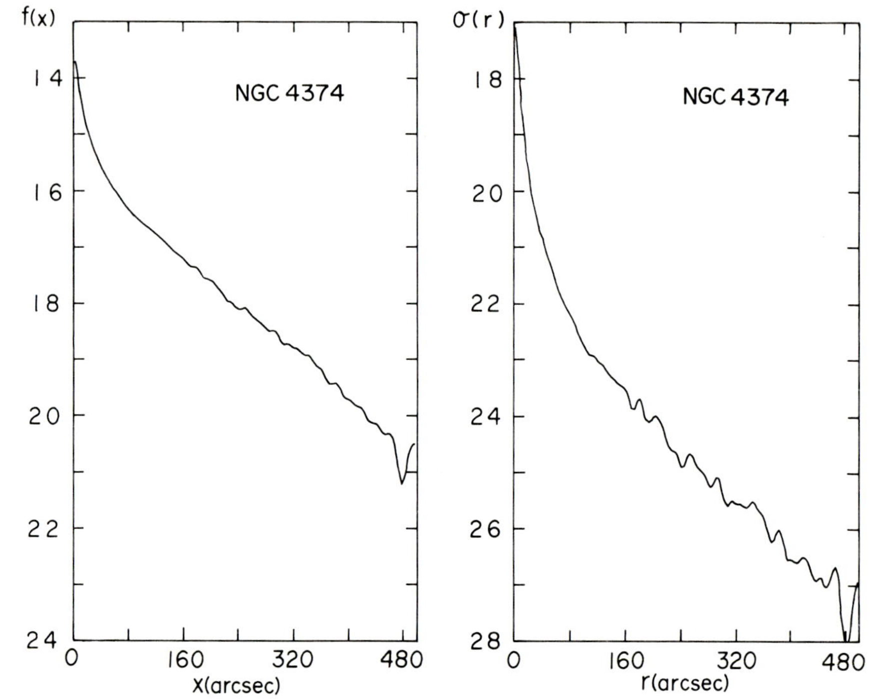


A5.1c: Projected profile and generalized radial profile

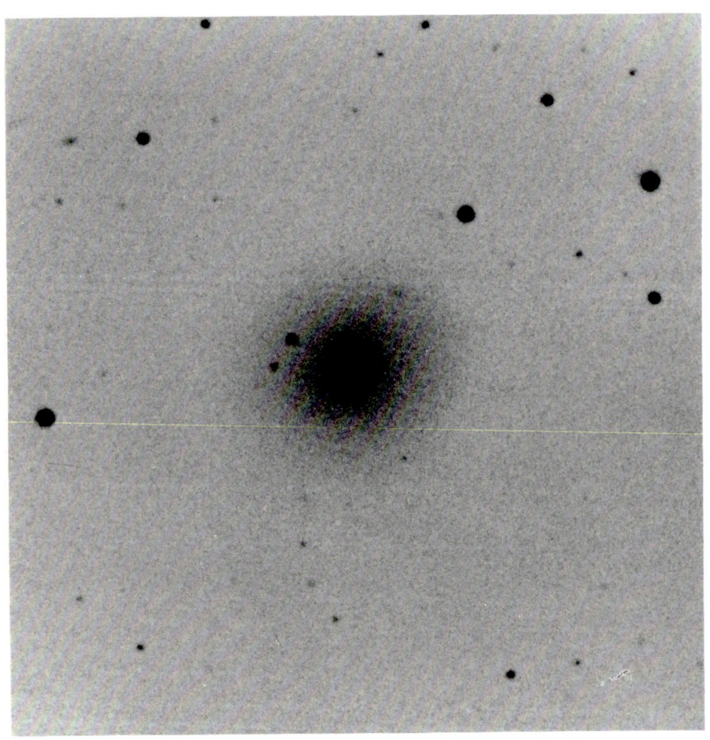

A5.2a: NGC 4267 SB(s)$0^-$? (V) $X1(P)$=1.82

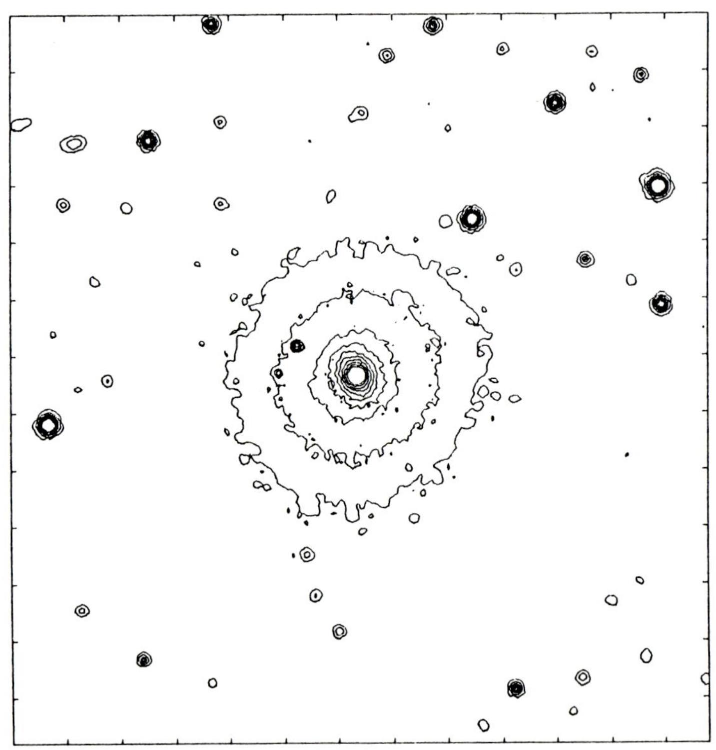

A5.2b: NGC 4267 SB(s)$0^-$? (V)

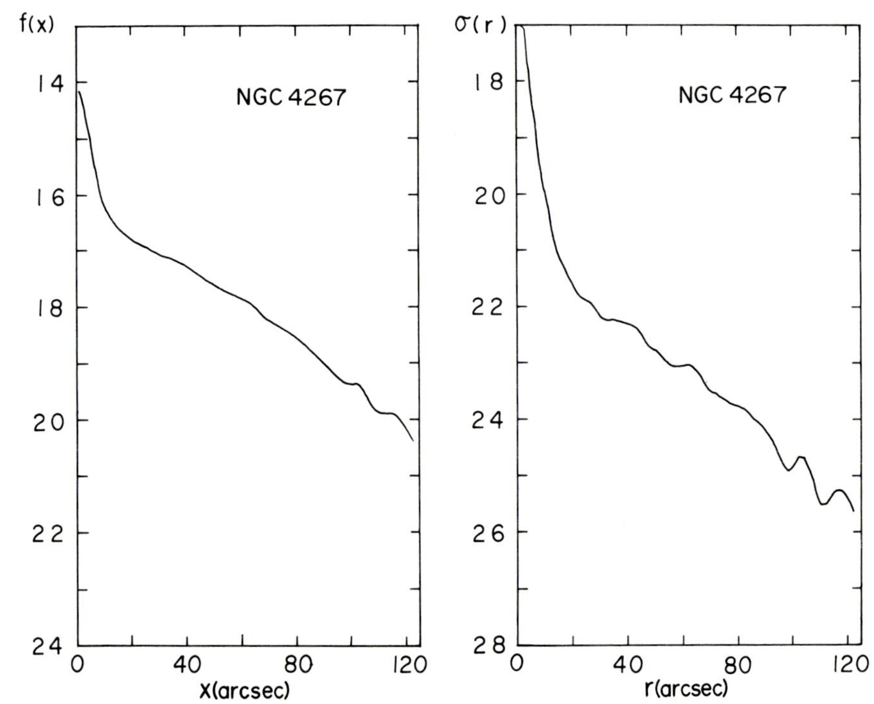


A5.2c: Projected profile and generalized radial profile

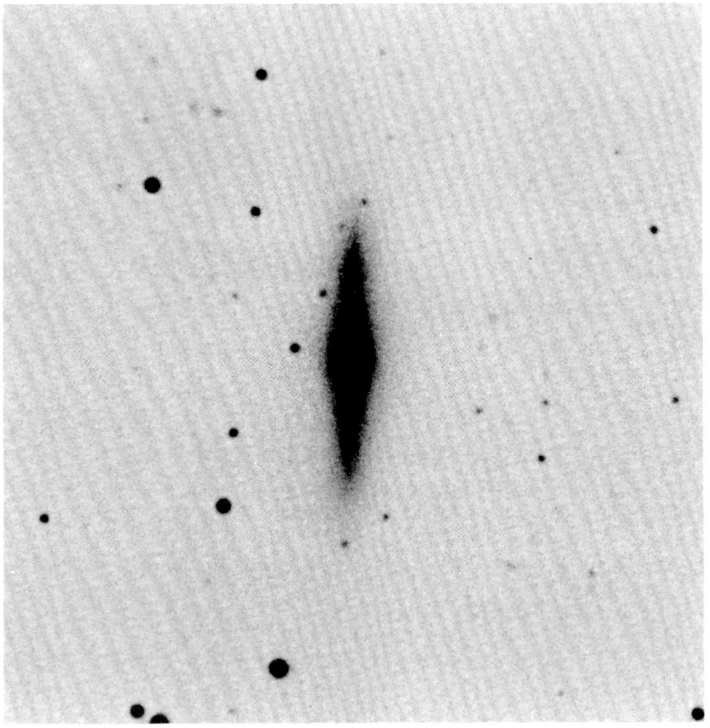

A5.3a: NGC 4026 S0sp (V) $X1(P)$=1.75

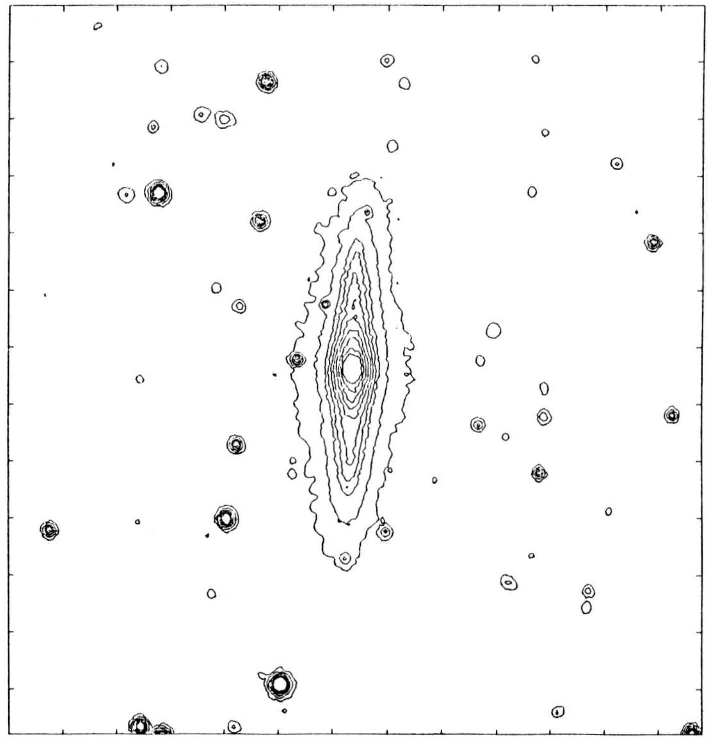

A5.3b: NGC 4026 S0sp (V)

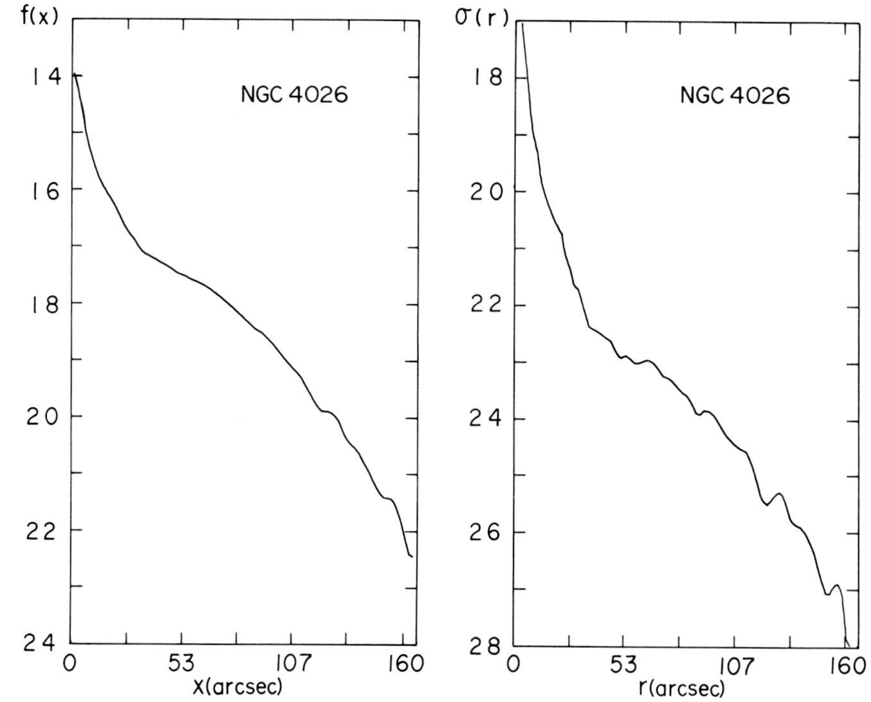


A5.3c: Projected profile and generalized radial profile

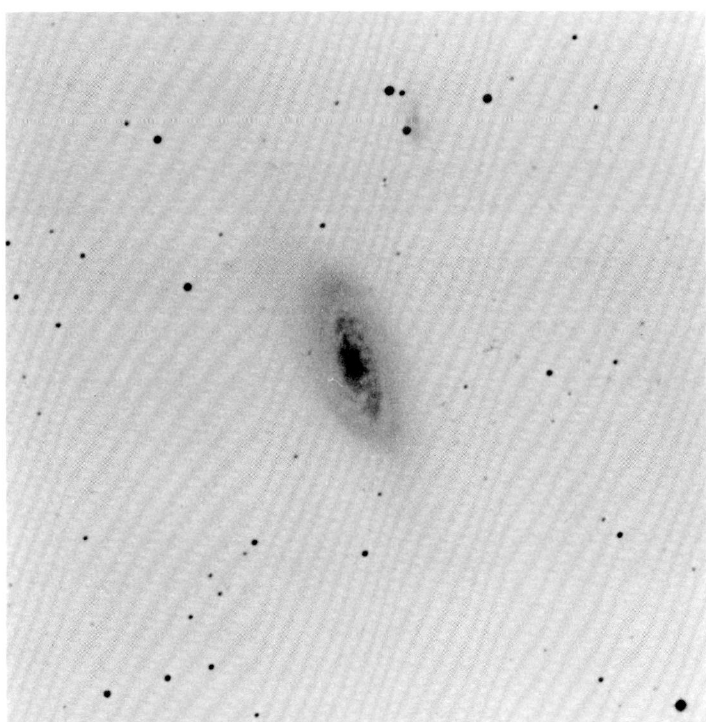

A5.4a: NGC 4569 SAB(rs)ab (V) $X1(P)$=0.21

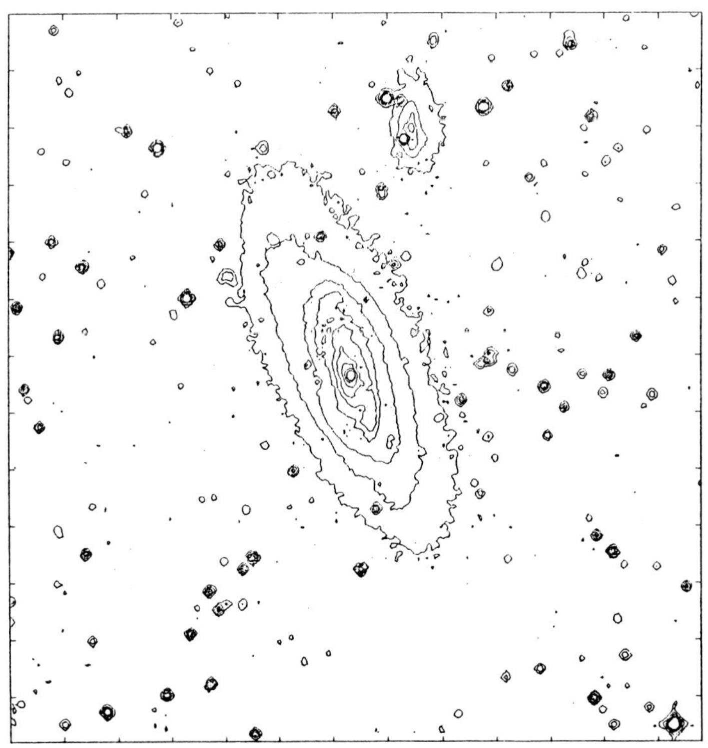

A5.4b: NGC 4569 SAB (rs)ab (V)

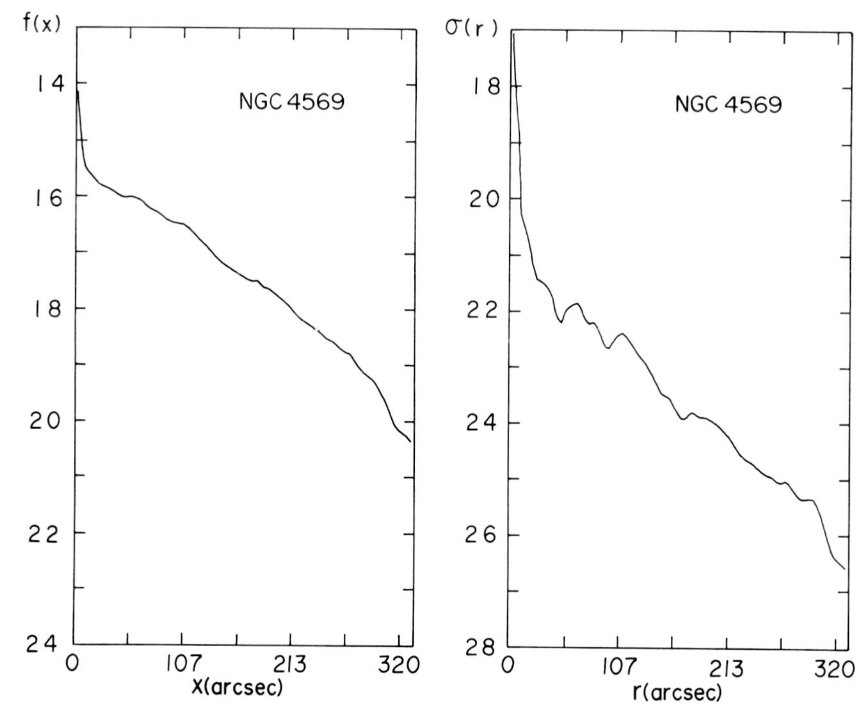


A5.4c: Projected profile and generalized radial profile

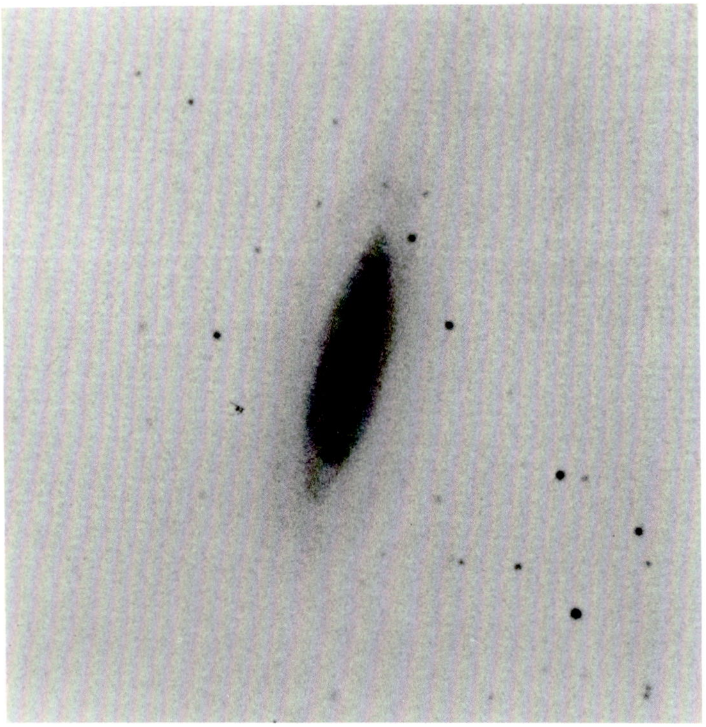

A5.5a: NGC 4100 (R')SA(rs)bc (V) $X1(P)$=-0.67

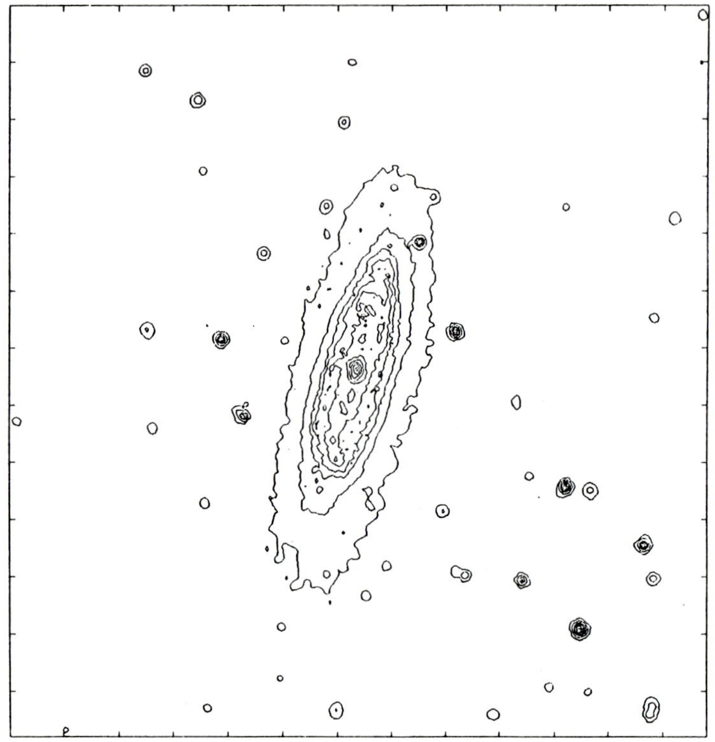

A5.5b: NGC 4100 (R')SA(rs)bc (V)

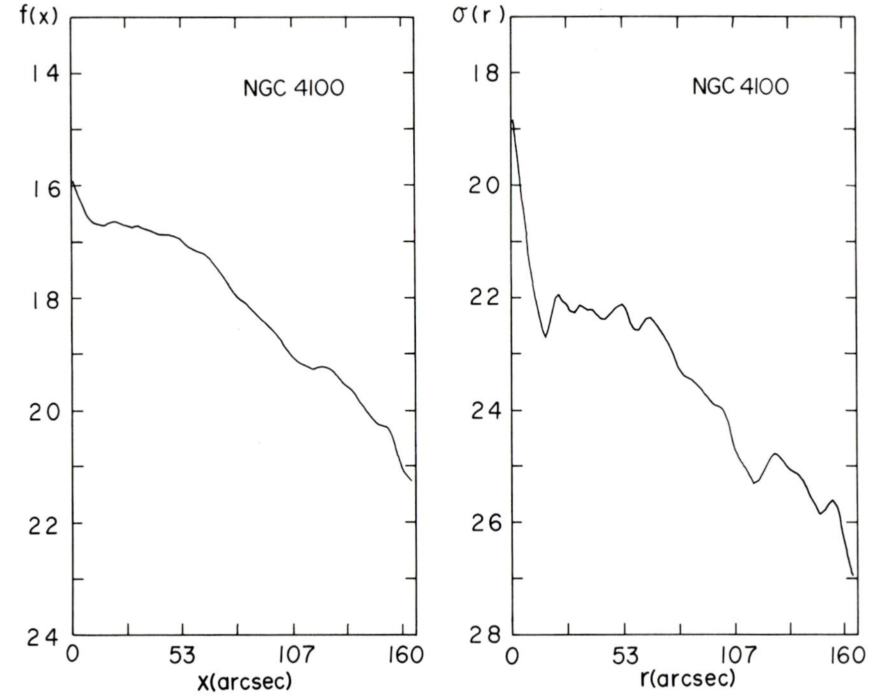


A5.5c: Projected profile and generalized radial profile

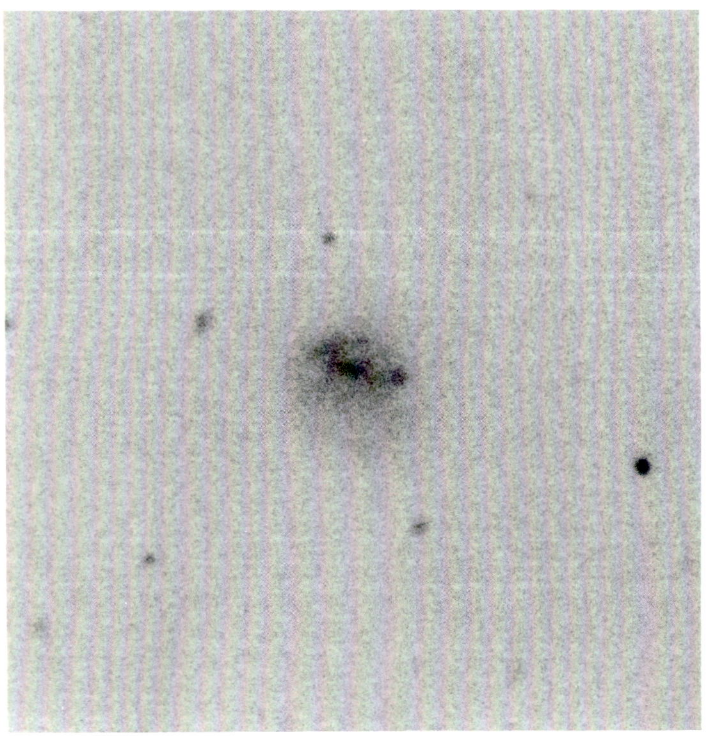

A5.6a: NGC 4234 SB(s)dm (V) $X1(P)$=-0.88

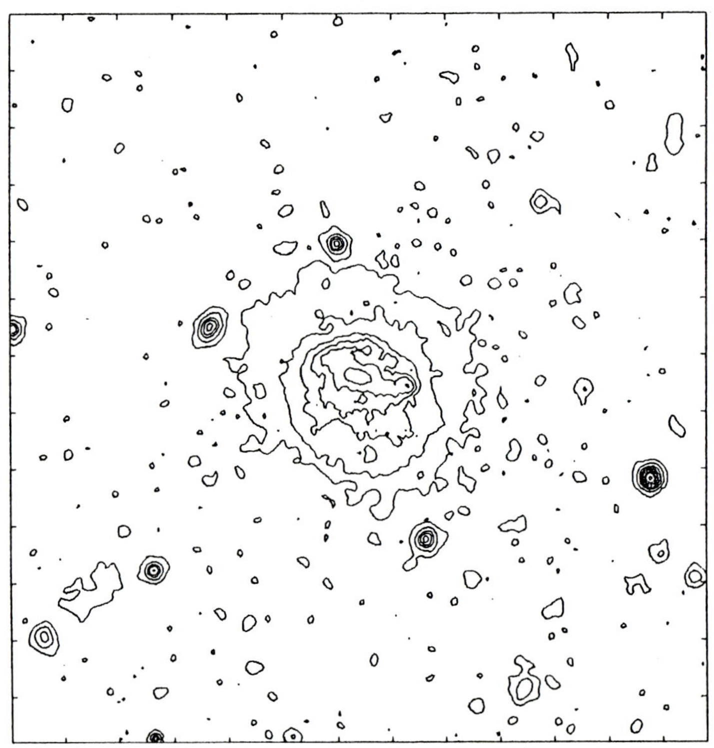

A5.6b: NGC 4234 SB(s)dm (V)

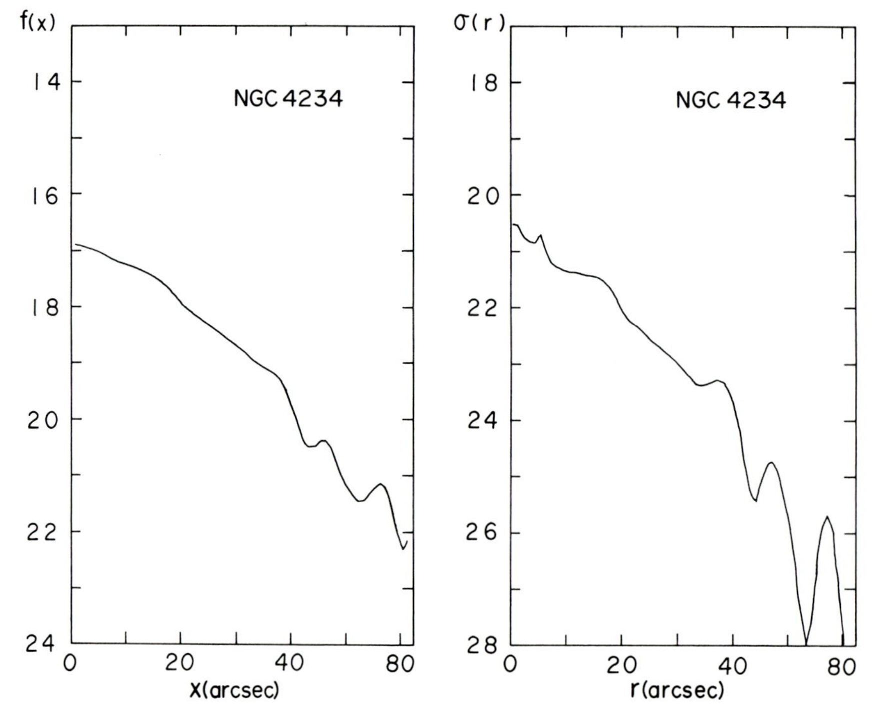


A5.6c: Projected profile and generalized radial profile

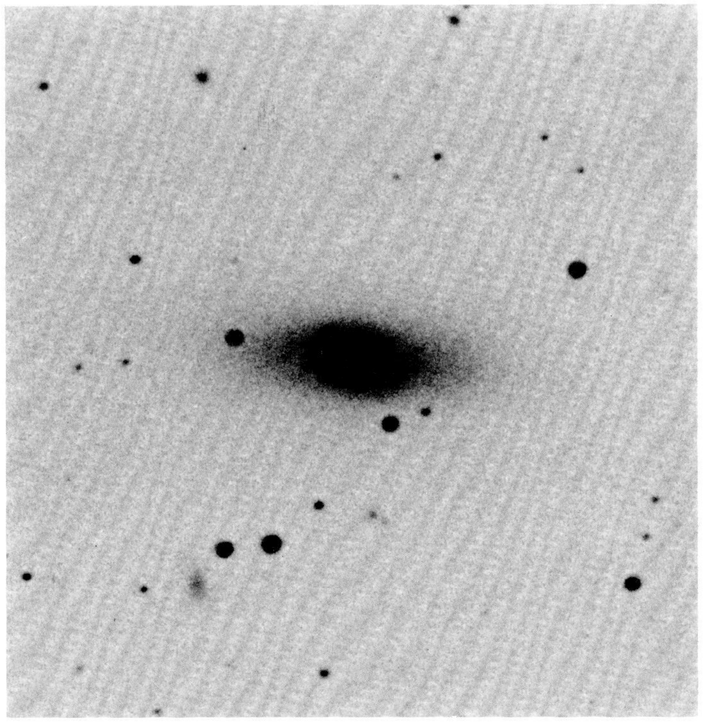

A5.7: NGC 4442 SB(s) $0^0$ (V)
$X1(P)$=1.80

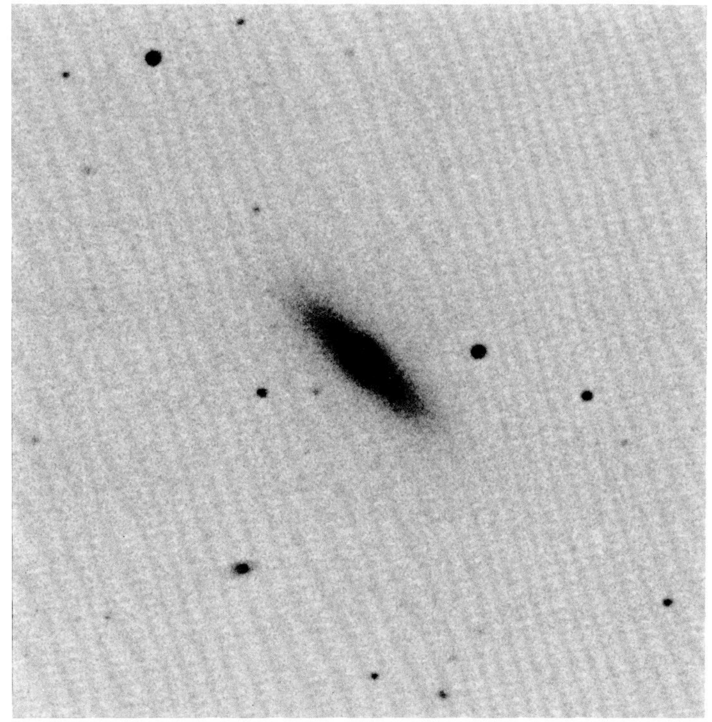

A5.8: NGC 4417 SB0:sp (V)
$X1(P)$=1.69

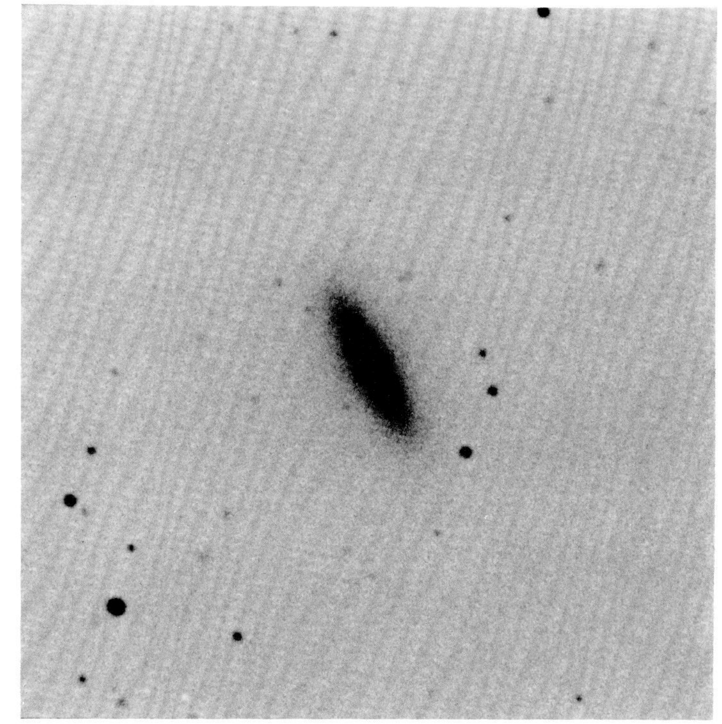

A5.9: NGC 4350 SA0sp (V)
$X1(P)$=1.61

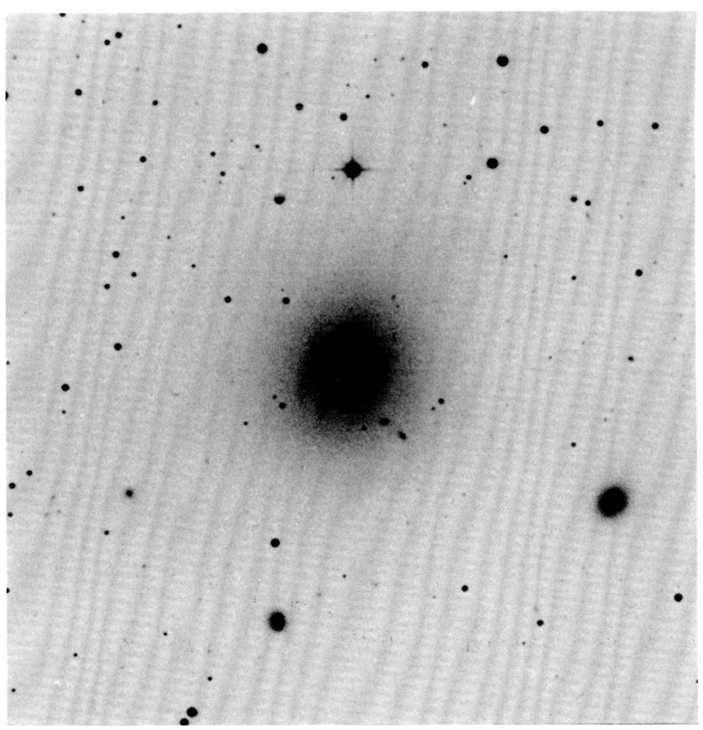

A5.10: NGC 4486 $E^+$0–1 P (V)
$X1(P)$=1.30

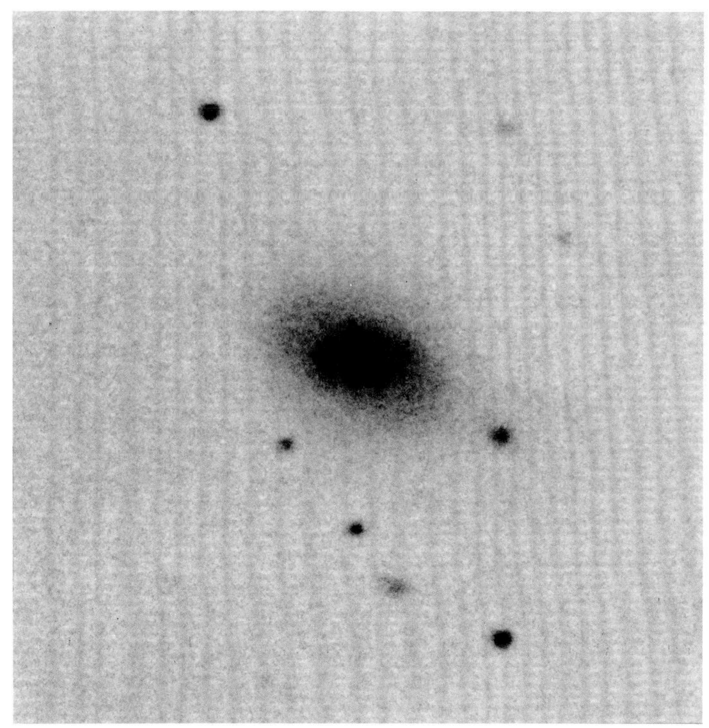

A5.11: NGC 4483 SB(s)0/a: (V)
$X1(P)$=1.29

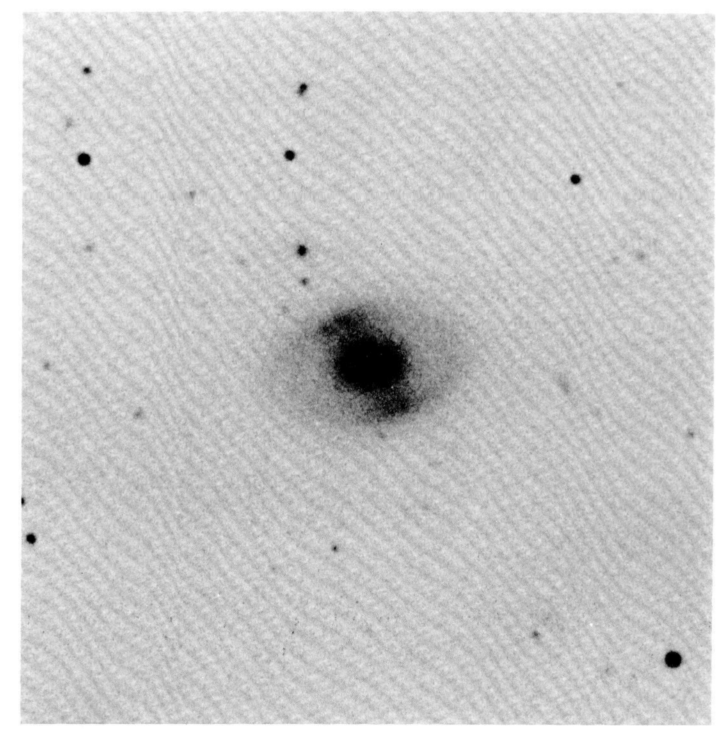

A5.12: NGC 4340 SB(r)$0^+$ (V)
$X1(P)$=1.28

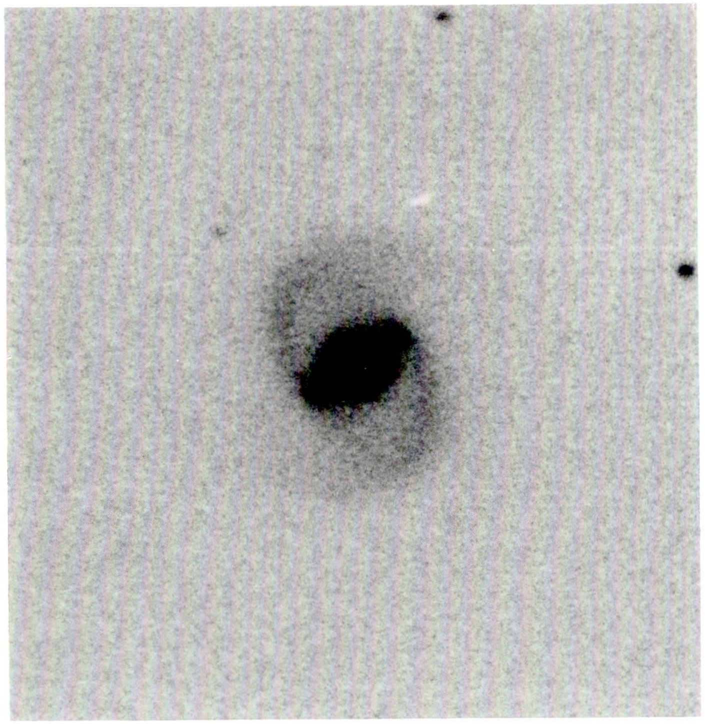

A5.13: NGC 4440 SB(rs)a (V)
$X1(P)$=1.27

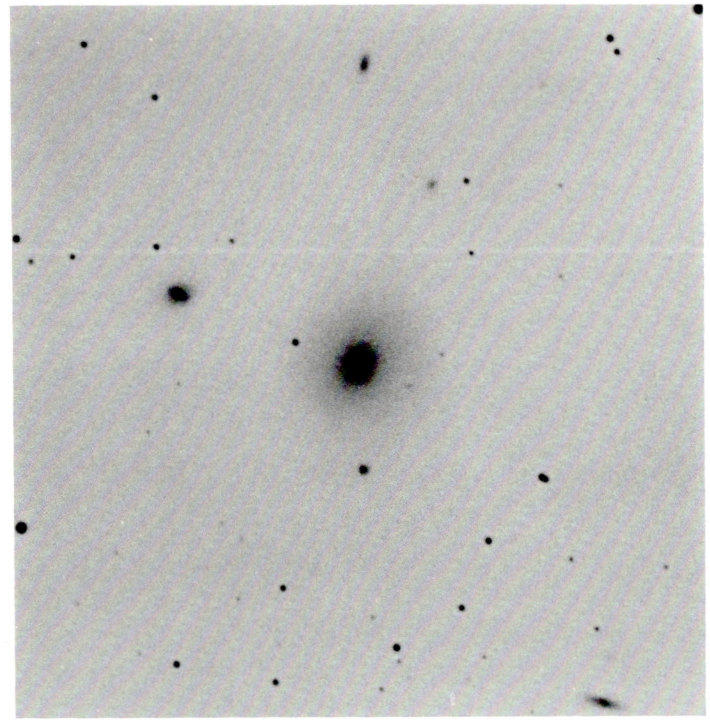

A5.14: NGC 4261 E2-3 (V)
$X1(P)$=1.15

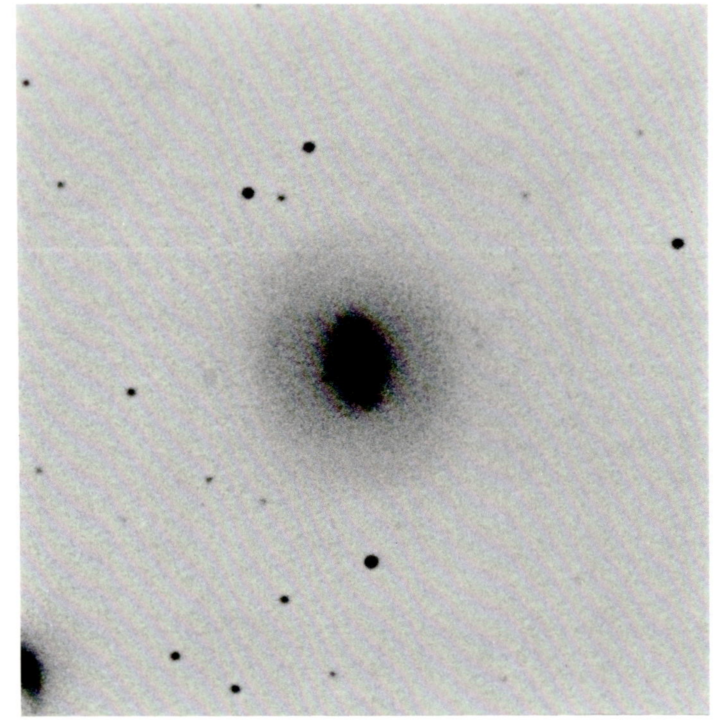

A5.15: NGC 4477 SB(s)0:? (V)
$X1(P)$=1.11

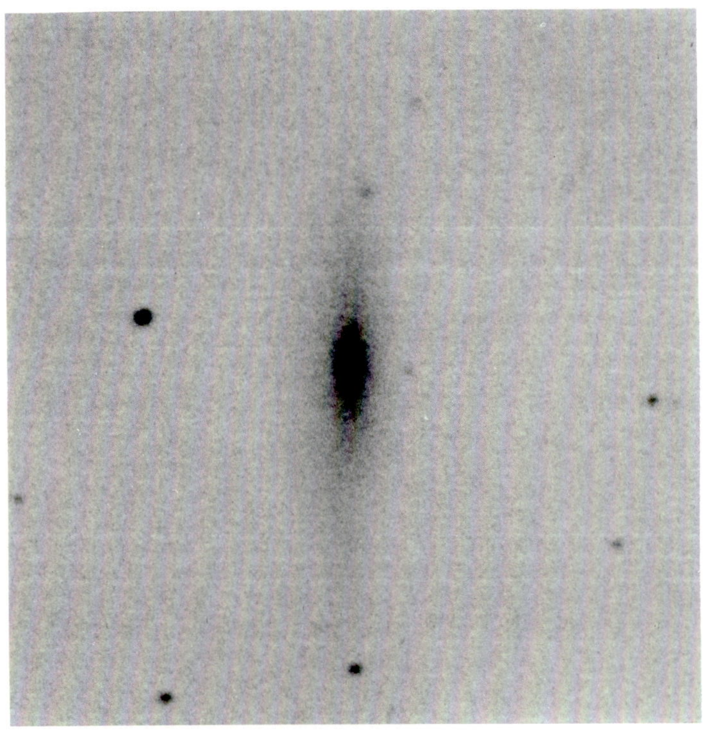

A5.16: NGC 4623 $SB0^{+}$?sp (V)
$X1(P)$=0.64

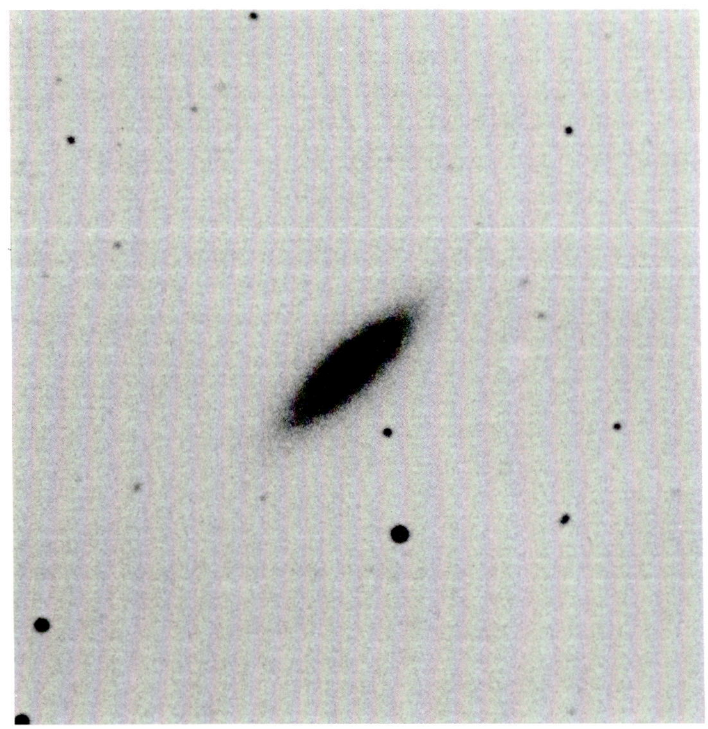

A5.17: NGC 4419 SB(s)a sp (V)
$X1(P)$=0.51

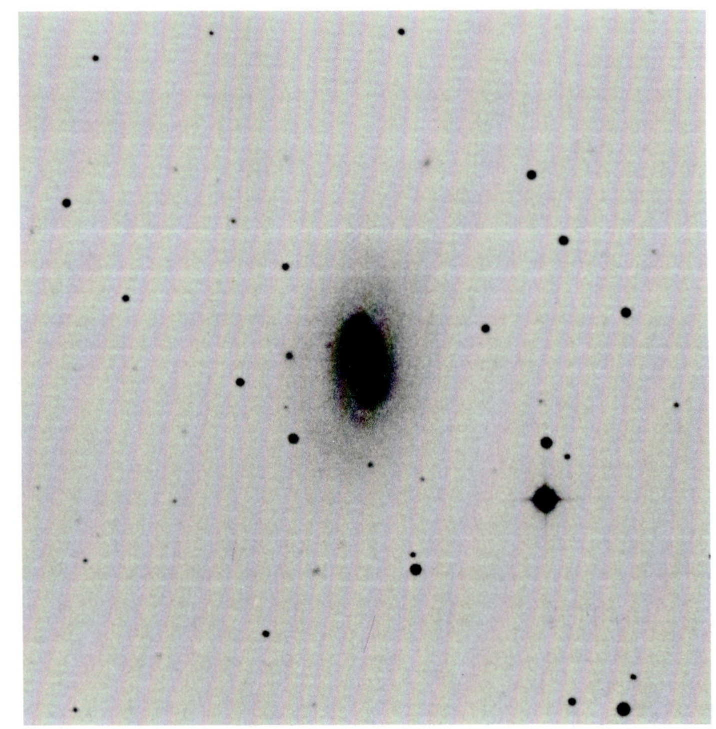

A5.18: NGC 4450 SA(s)ab (V)
$X1(P)$=0.47

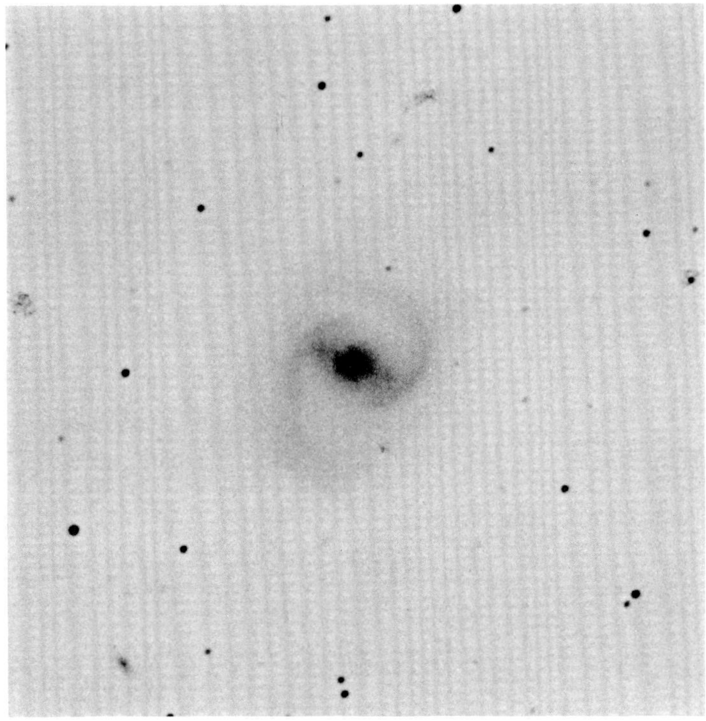

A5.19: NGC 4548 SB(rs)b (V)
$X1(P)$=0.10

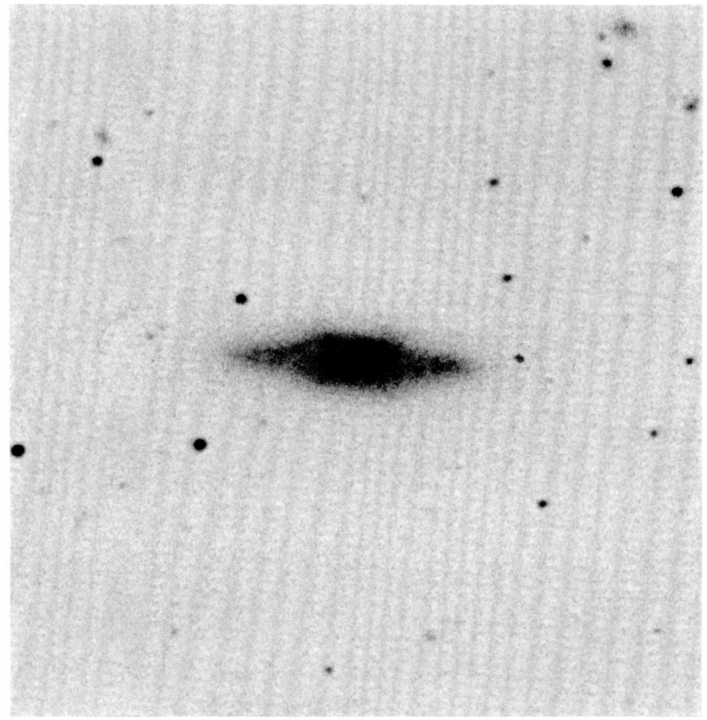

A5.20: NGC 4469 SB(s)0/a?sp (V)
$X1(P)$=0.08

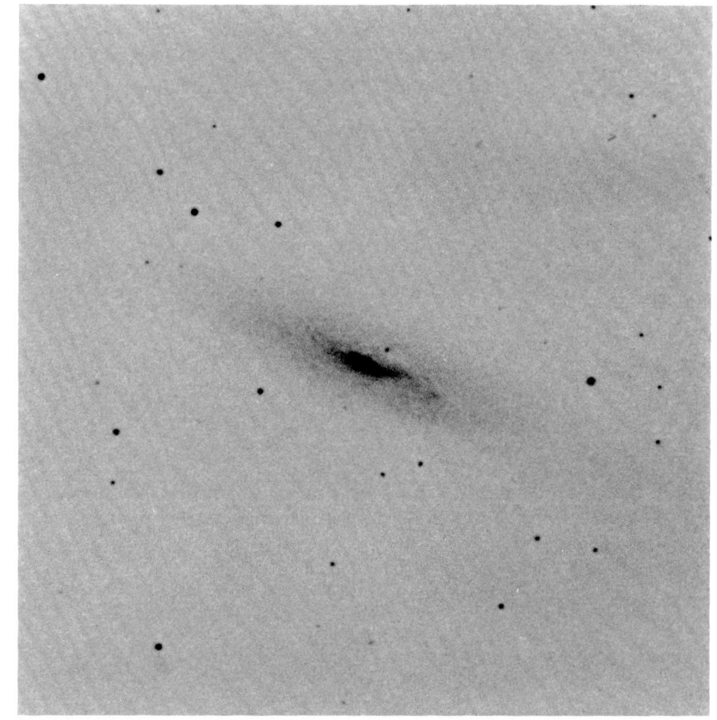

A5.21: NGC 4527 SAB(s)bc (V)
$X1(P)$=-0.07

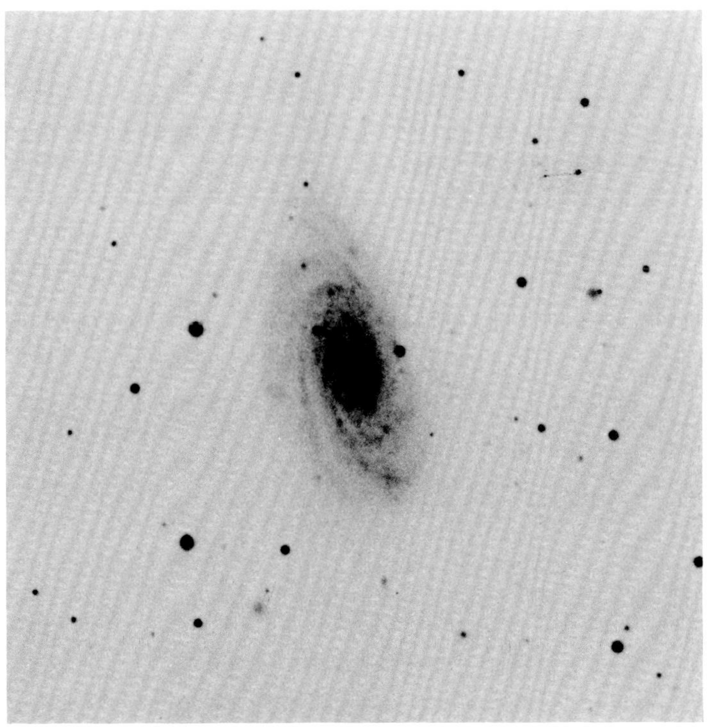

A5.22: NGC 3953 SB(r)bc (V)
$X1(P)$=-0.09

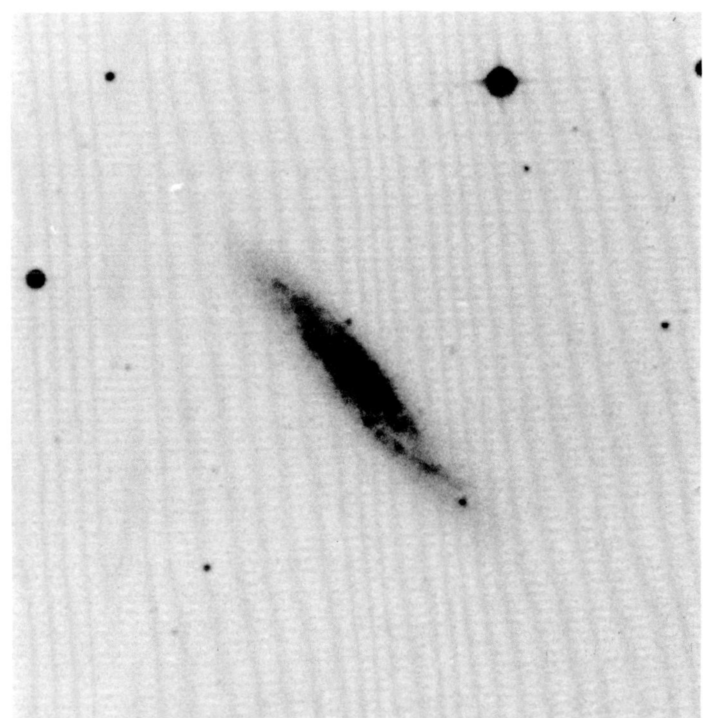

A5.23: NGC 3877 SA(s)c: (V)
$X1(P)$=-0.39

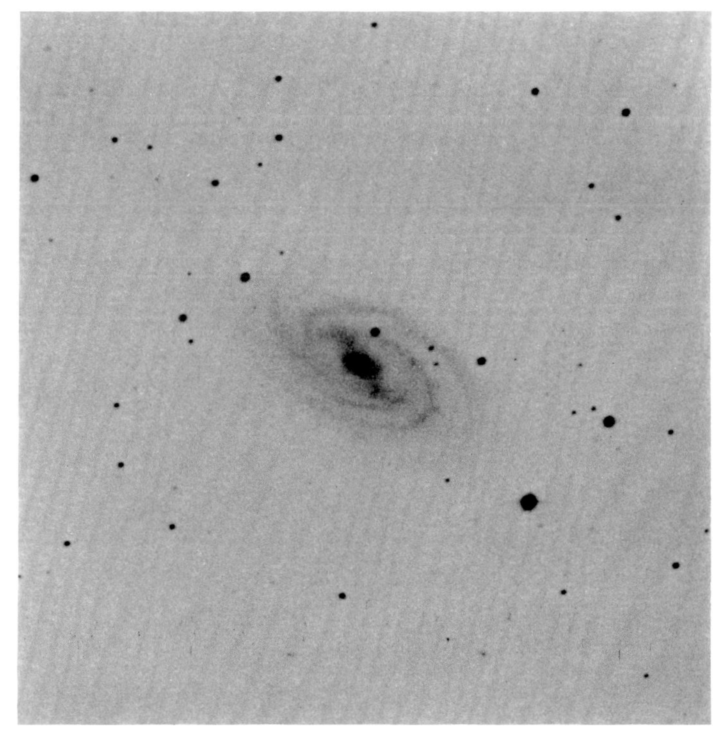

A5.24: NGC 3992 SB(rs)bc (V)
$X1(P)$=-0.43

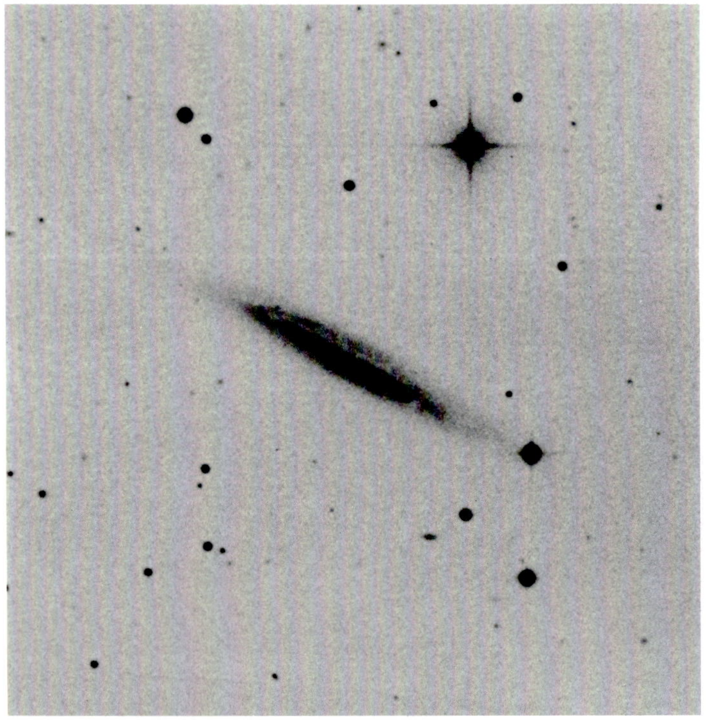

A5.25: NGC 4157 SAB(s)b?sp (V)
X1(P)=-0.43

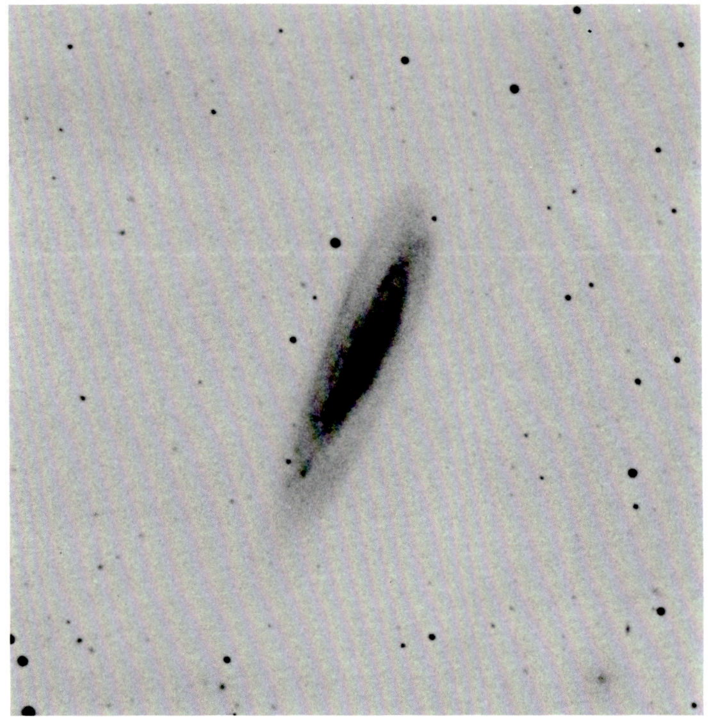

A5.26: NGC 4192 SAB(s)ab (V)
X1(P)=-0.44

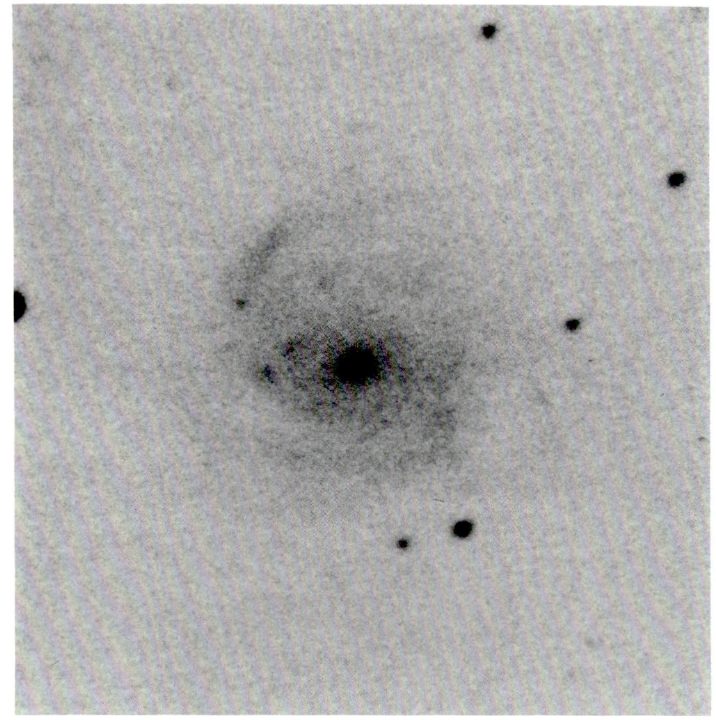

A5.27: NGC 3780 SA(s)c: (V)
X1(P)=-0.75

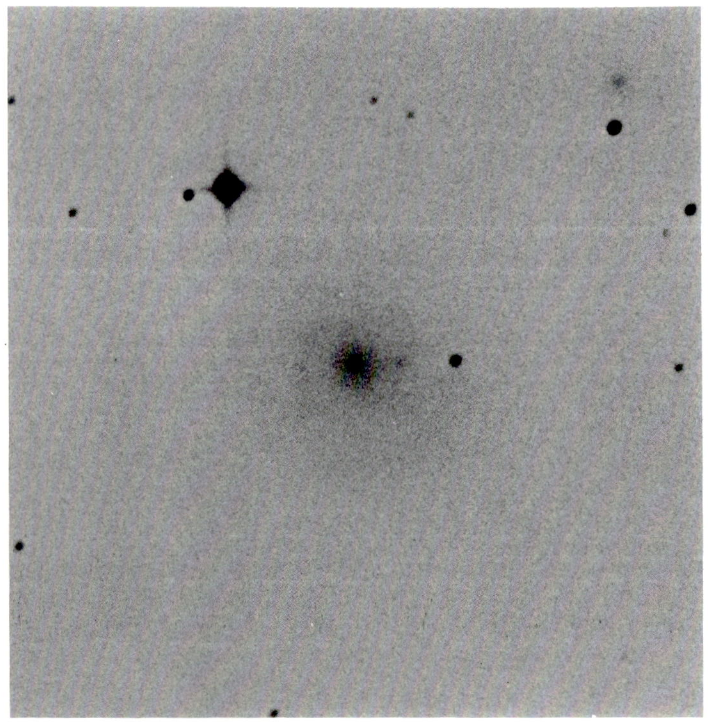

A5.28: NGC 4571 SA(r)d (V)
X1(P)=-0.77

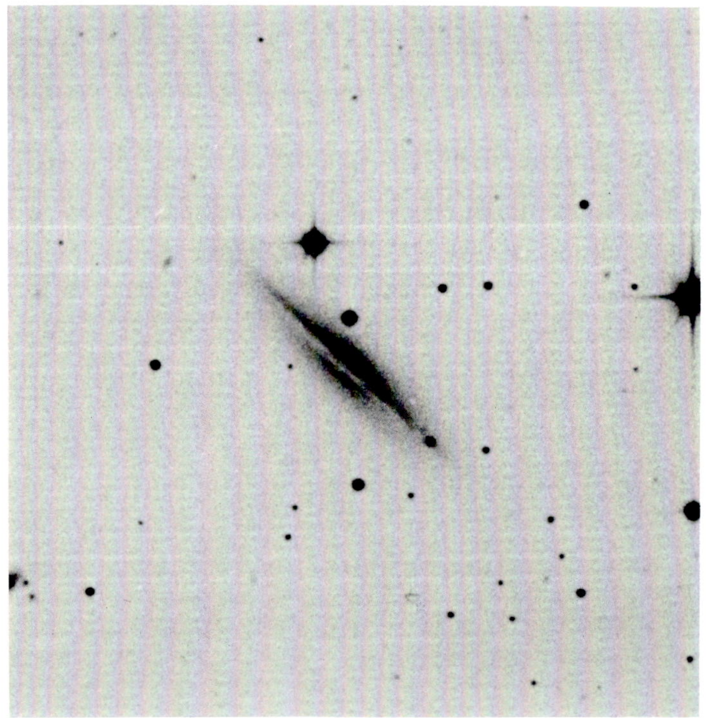

A5.29: NGC 4217 Sb sp (V)
X1(P)=-0.85

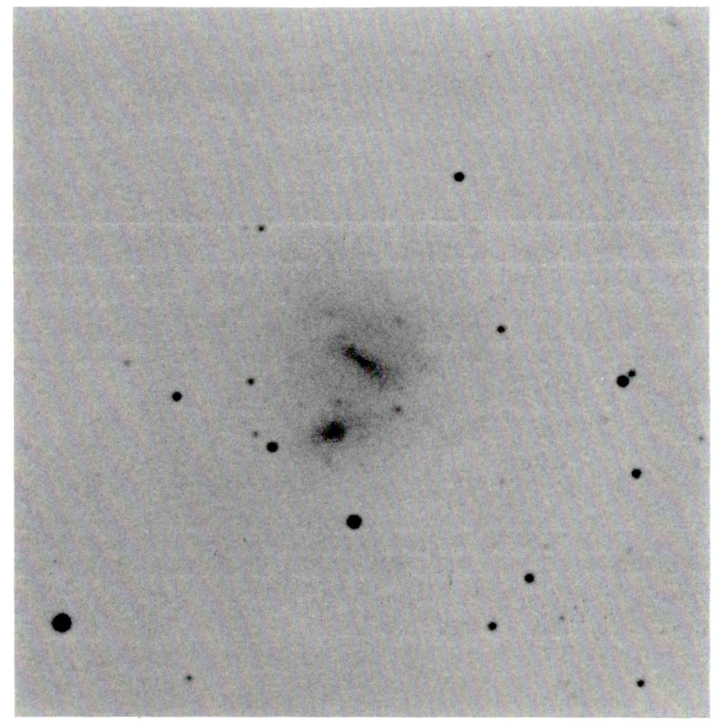

A5.30: NGC 4496A SB(rs)m (V)
X1(P)=-1.09

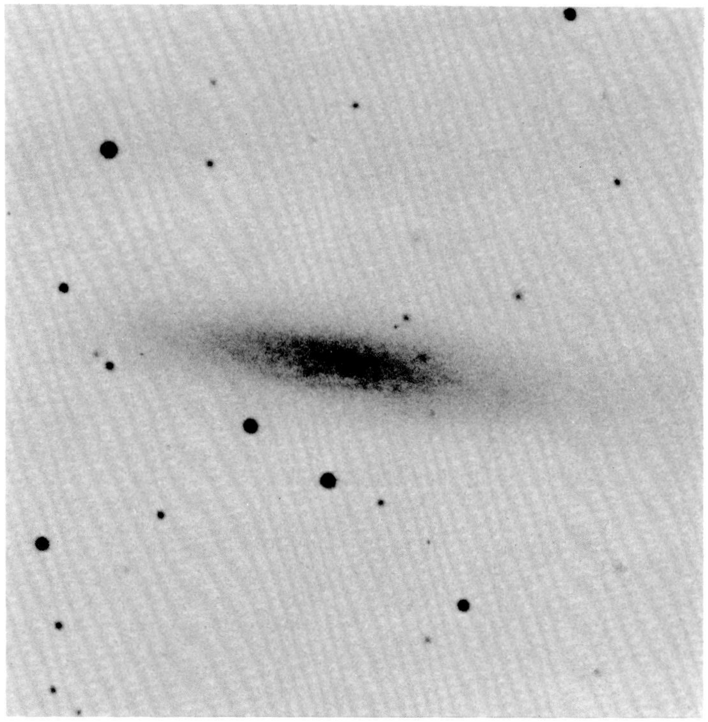

A5.31: NGC 3917 SAcd: (V)
*X1*(*P*)=-1.12

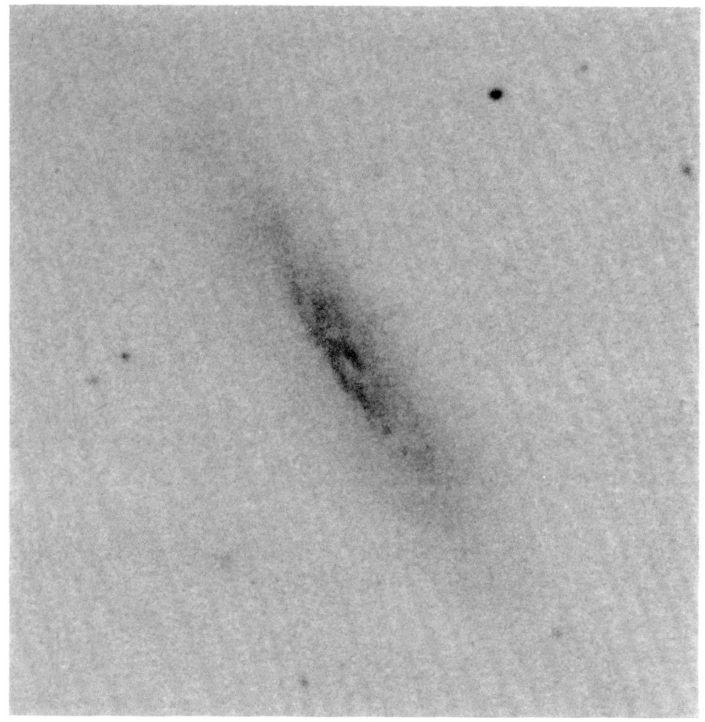

A5.32: NGC 4522 SB(s)cd:sp (V)
*X1*(*P*)=-1.23

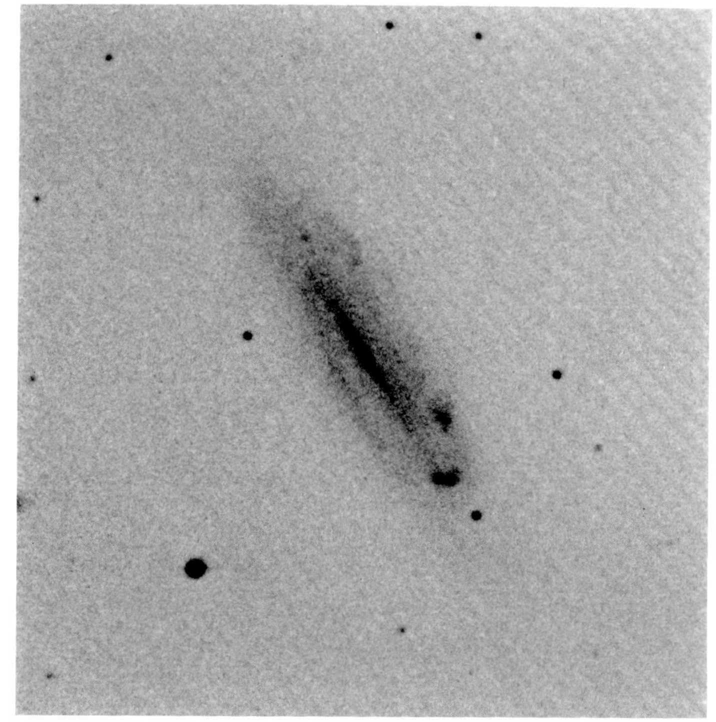

A5.33: NGC 4178 SB(rs)dm (V)
*X1*(*P*)=-1.27

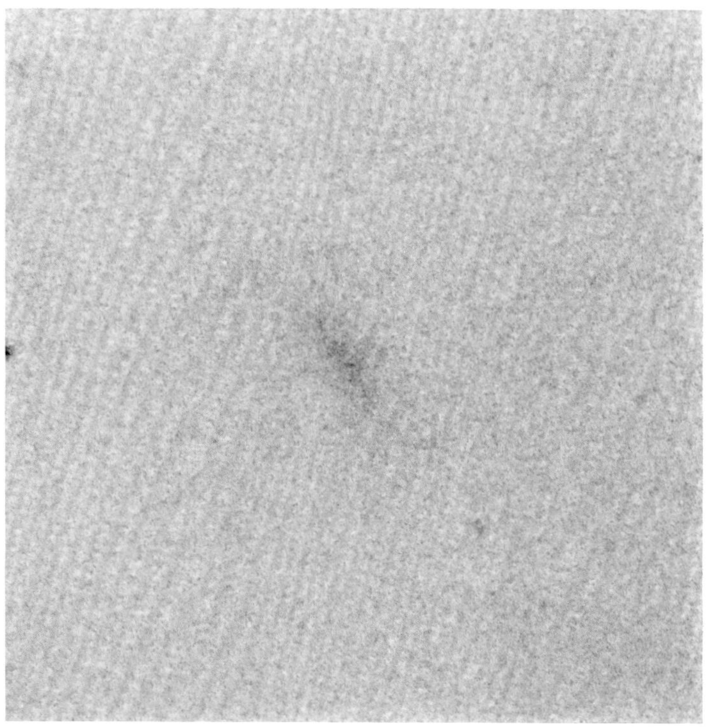

A5.34: A1155+51 SB(s)m (V)
*X1*(*P*)=-1.41

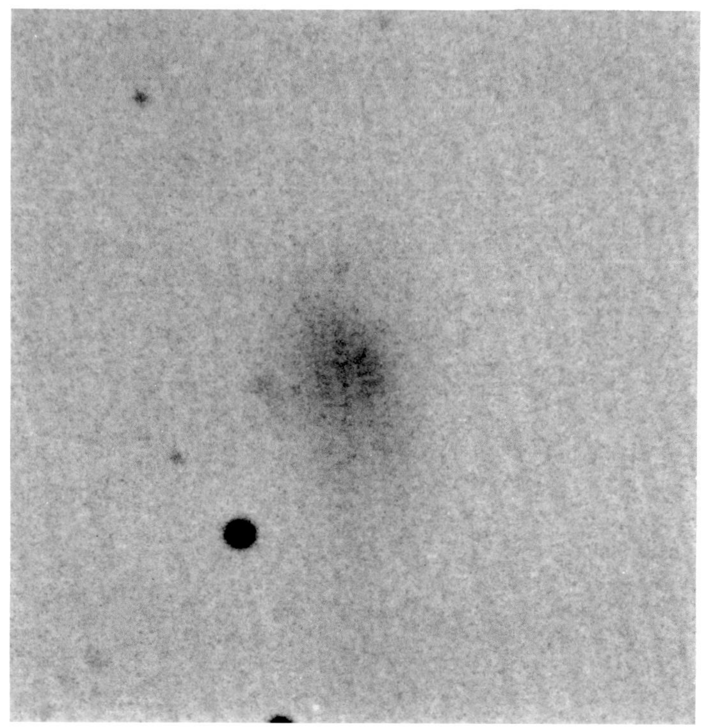

A5.35: IC3059 Im (V)
*X1*(*P*)=-1.44

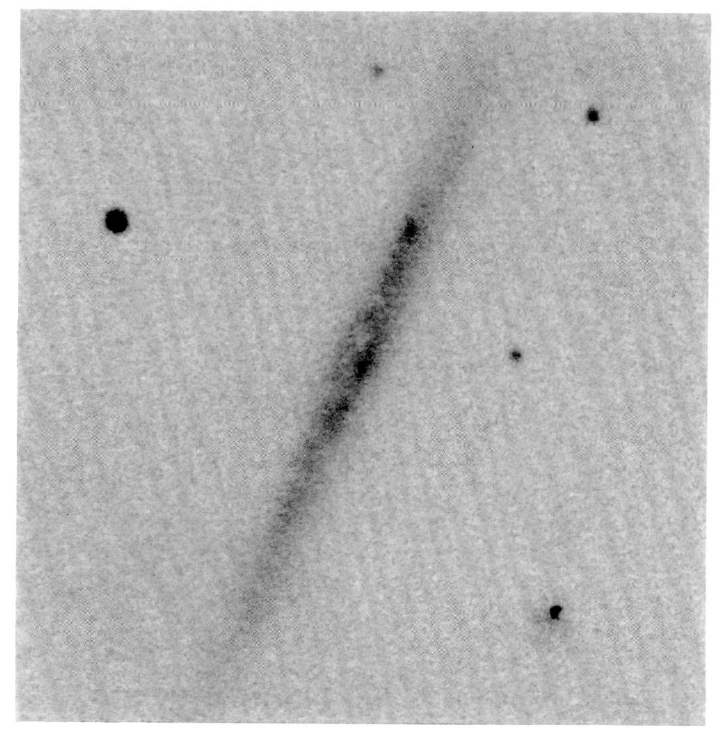

A5.36: IC3322A SB(s)cd:sp (V)
*X1*(*P*)=-1.46

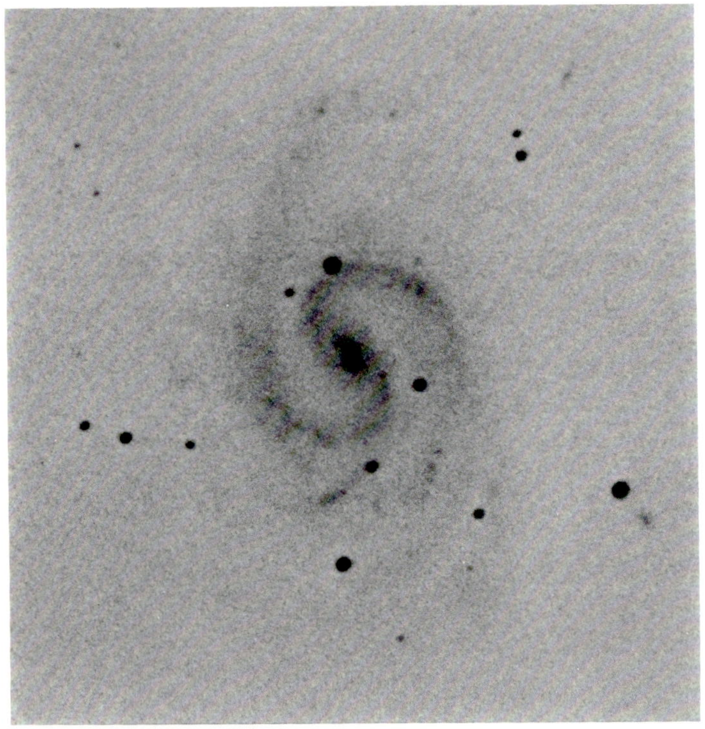

A6.1a: NGC 4535 SAB(s)c (V)

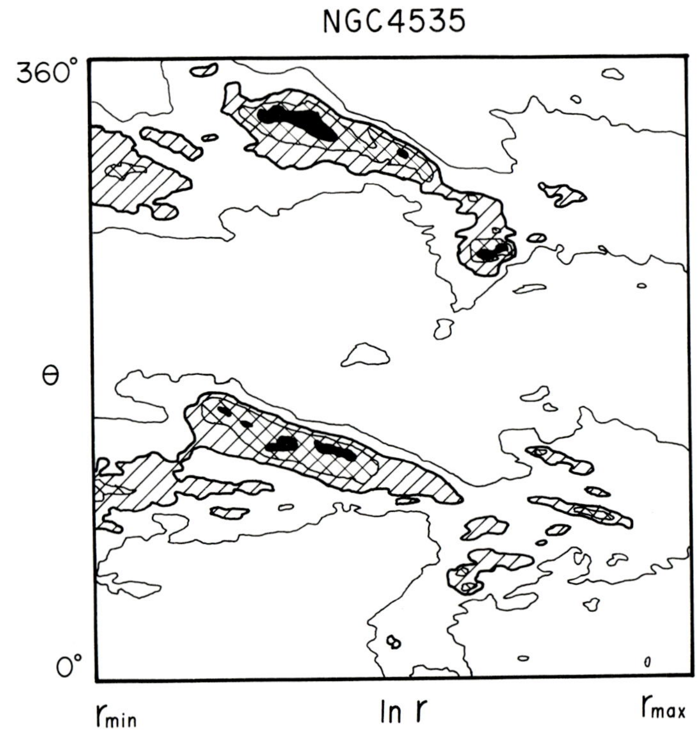


A6.1b: Face-on intensity distribution

A6.1c: Reconstructed face-on image

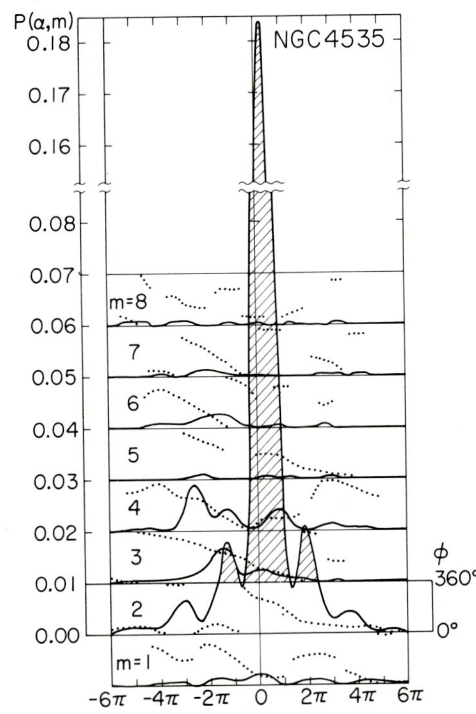


A6.1d: Normalized power spectrum

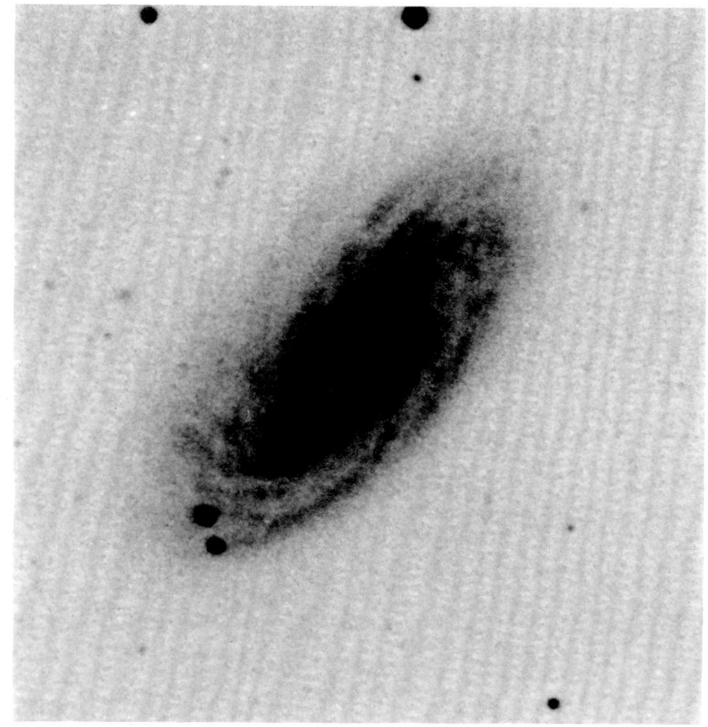

A6.2a: NGC 4501 SA(rs)b (V)

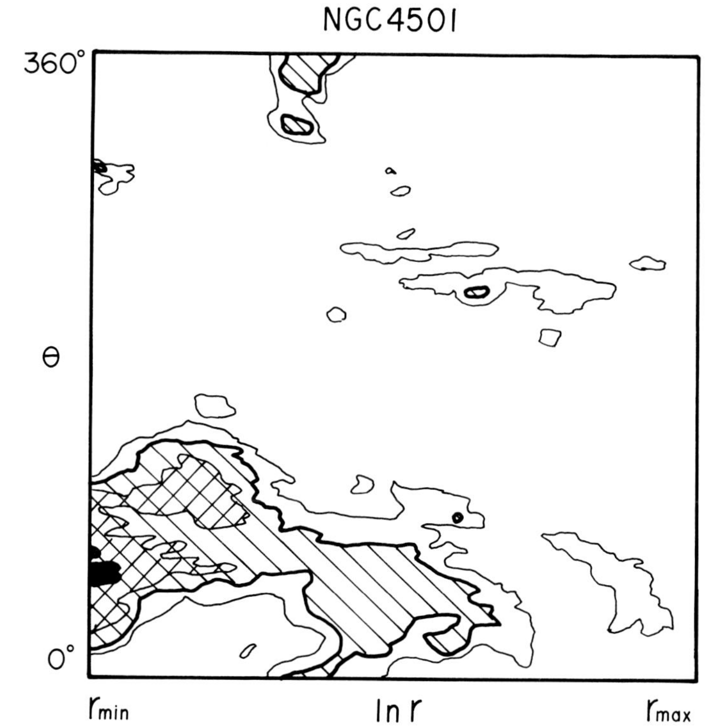


A6.2b: Face-on intensity distribution

A6.2c: Reconstructed face-on image

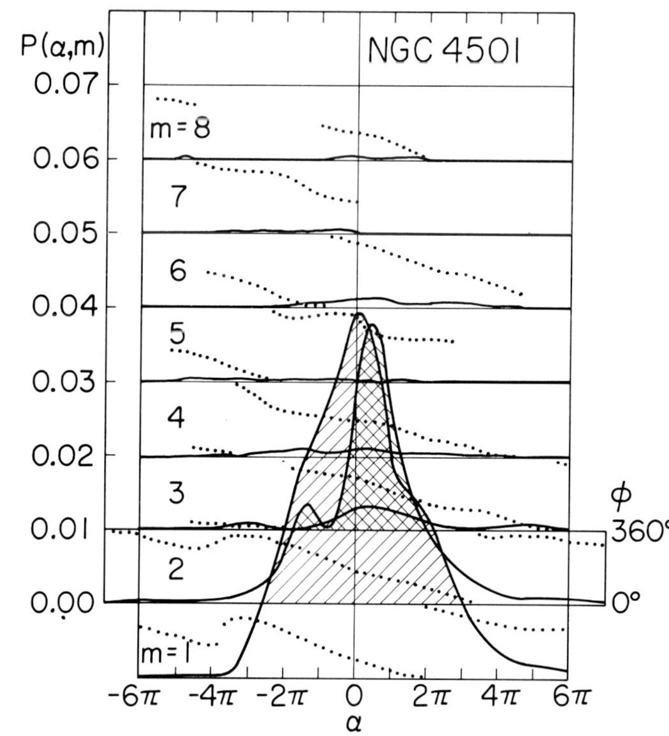


A6.2d: Normalized power spectrum

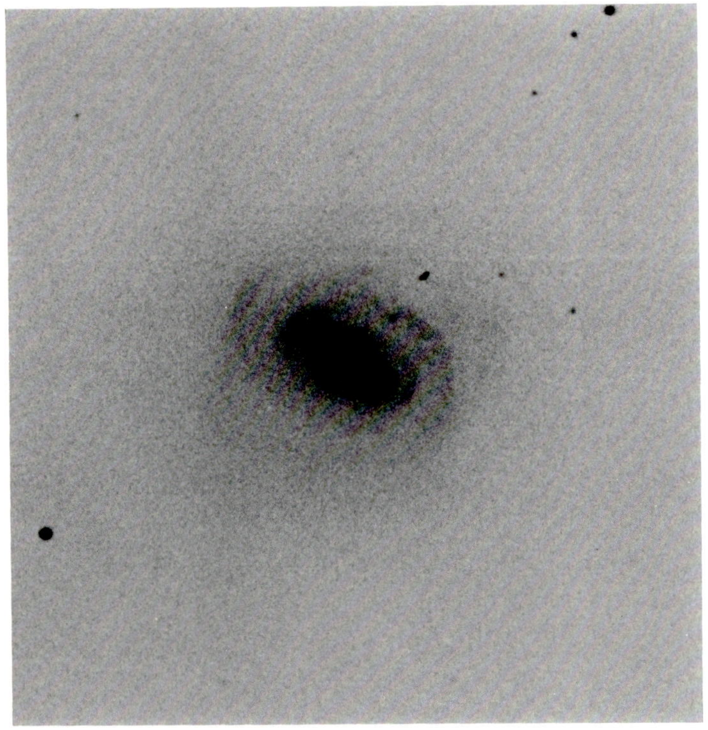

A6.3a: NGC 4579 SAB(rs)b (V)

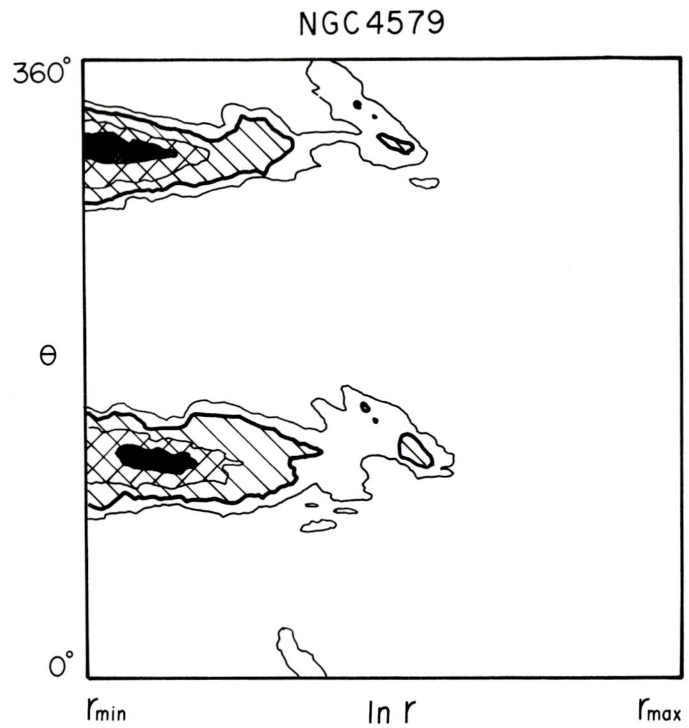


A6.3b: Face-on intensity distribution

A6.3c: Reconstructed face-on image

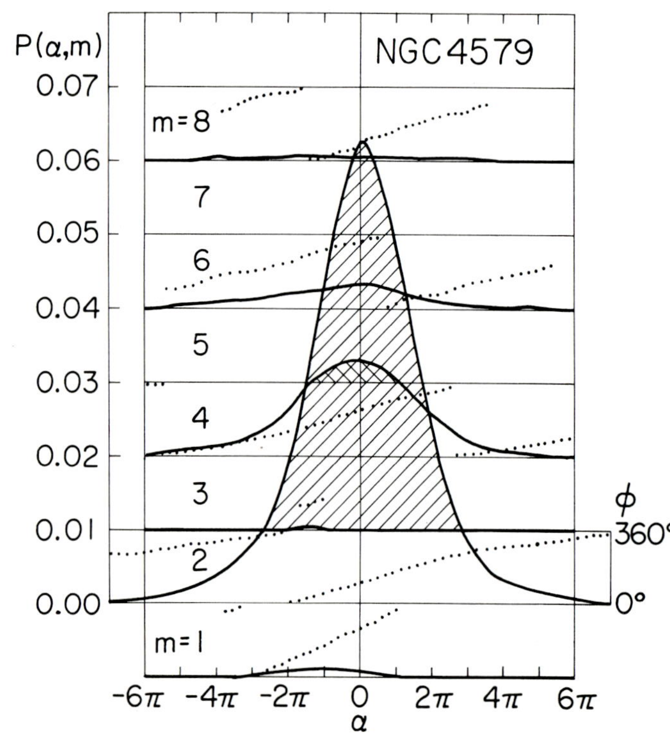


A6.3d: Normalized power spectrum

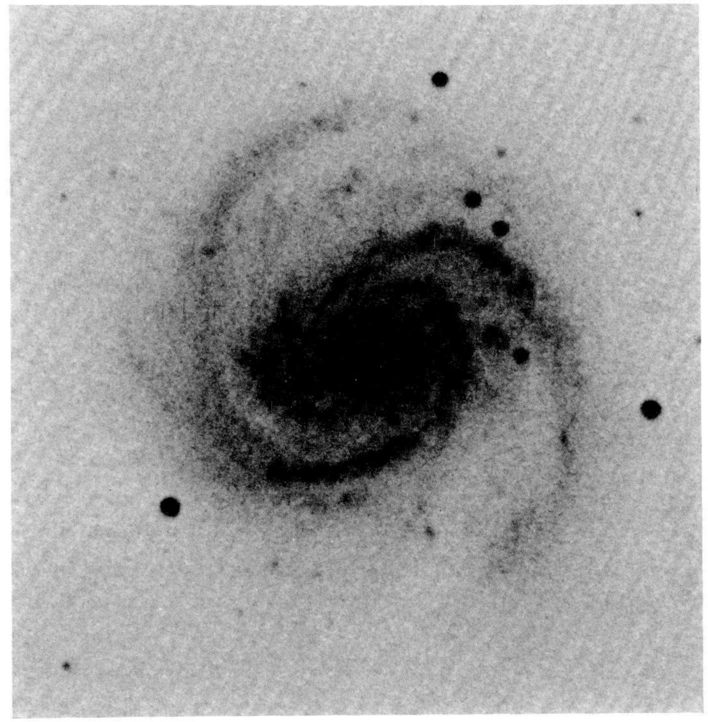

A6.4a: NGC 4321 SAB(s)bc (V)

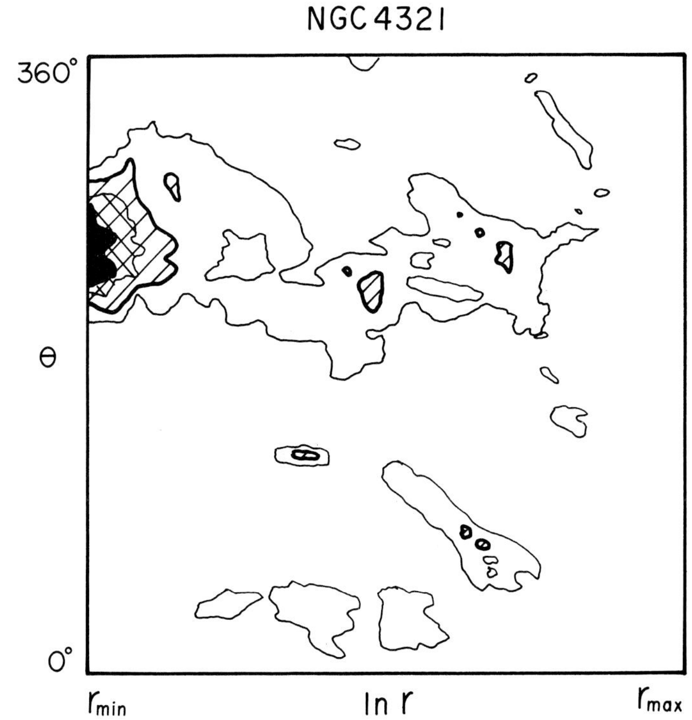


A6.4b: Face-on intensity distribution

A6.4c: Reconstructed face-on image

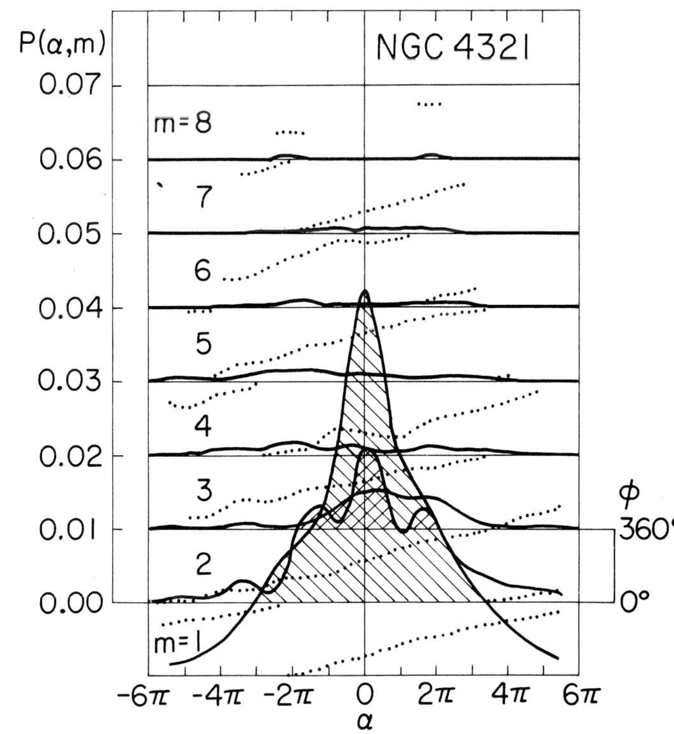


A6.4d: Normalized power spectrum

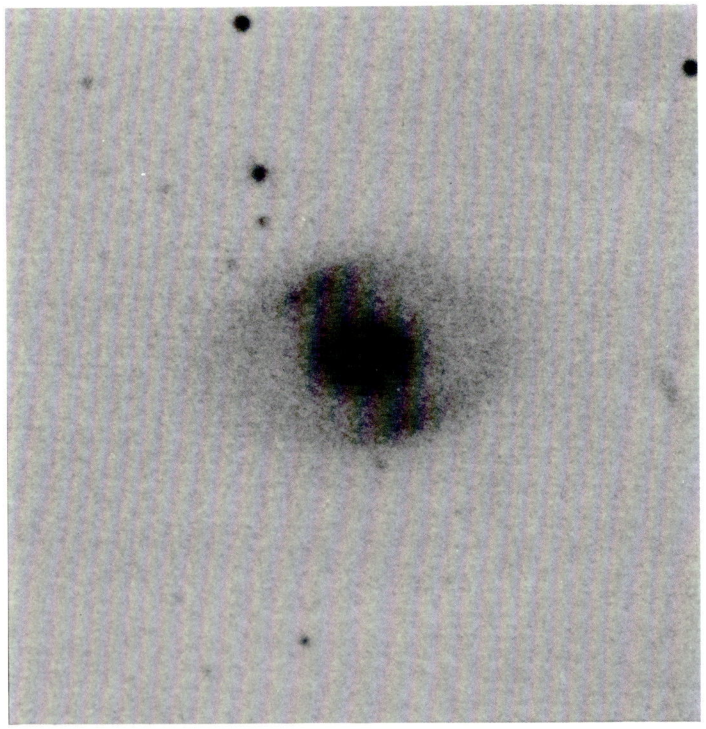

A6.5a: NGC 4340 SB(r)0⁺ (V)

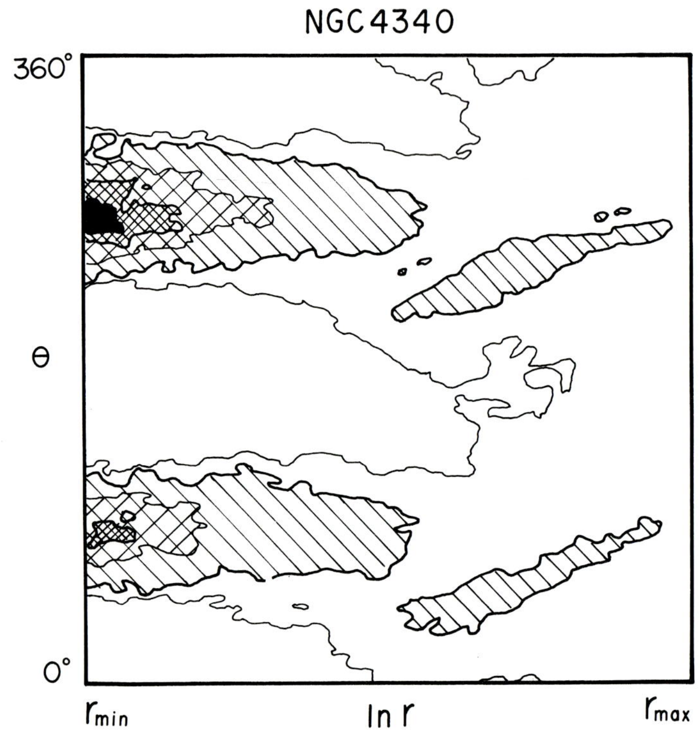


A6.5b: Face-on intensity distribution

A6.5c: Reconstructed face-on image

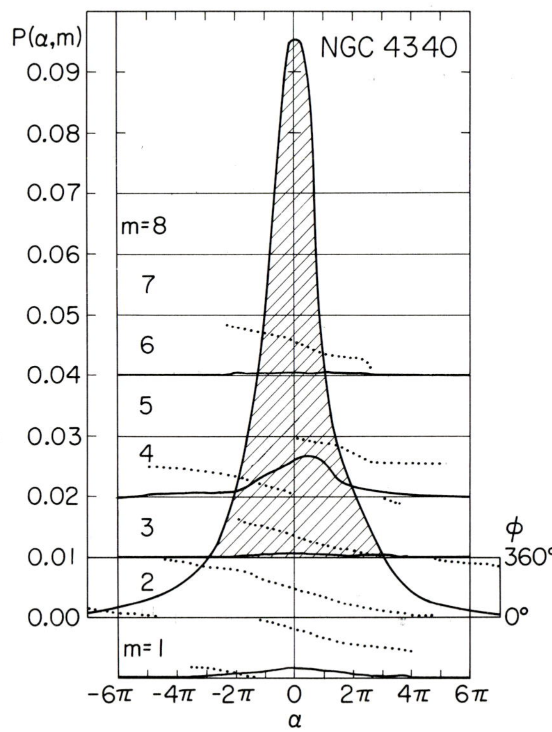


A6.5d: Normalized power spectrum

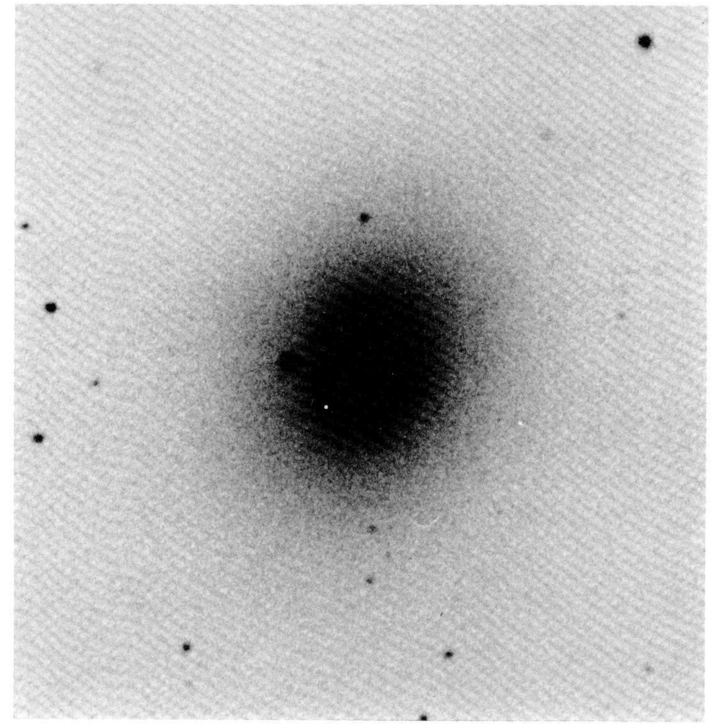

A6.6a: NGC 4472 E2 (V)

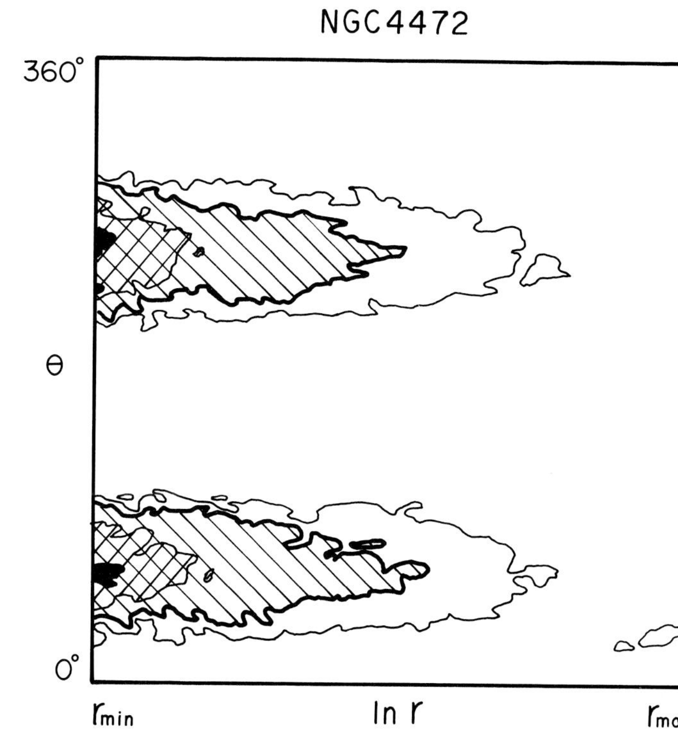


A6.6b: Face-on intensity distribution

A6.6c: Reconstructed face-on image

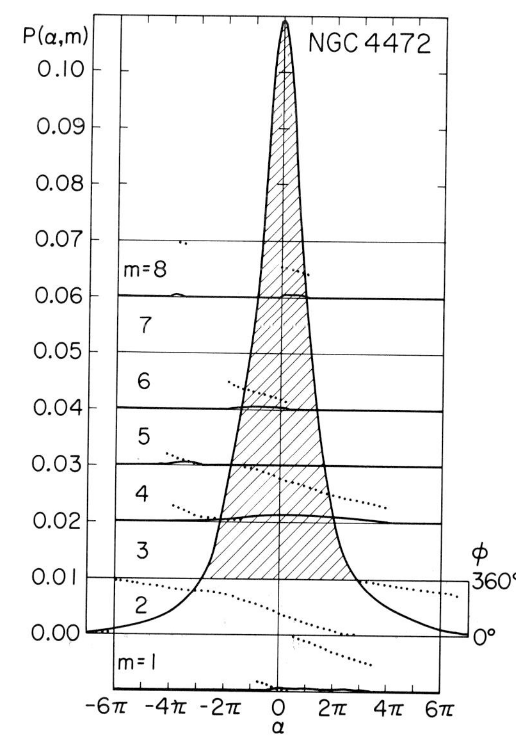


A6.6d: Normalized power image

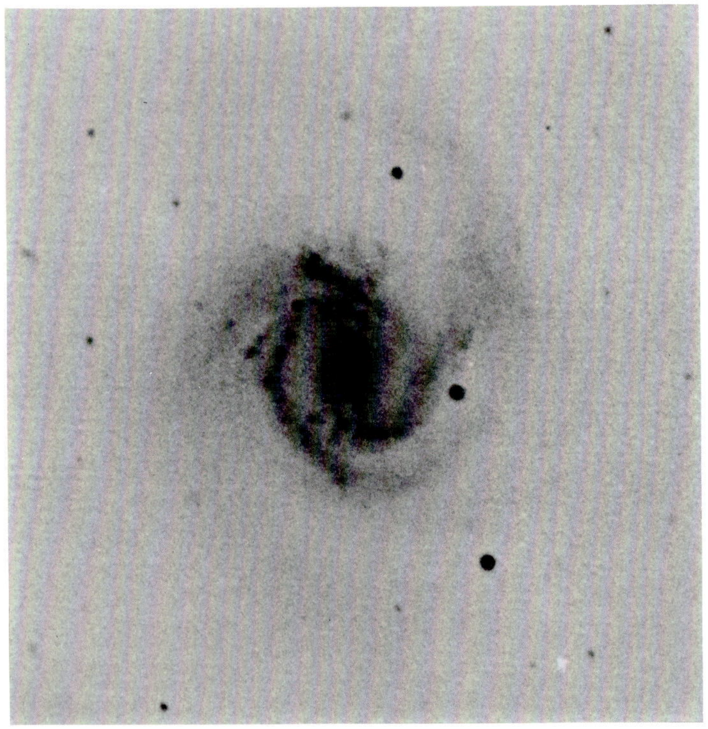

A6.7a: NGC 4303 SAB(rs)bc (V)

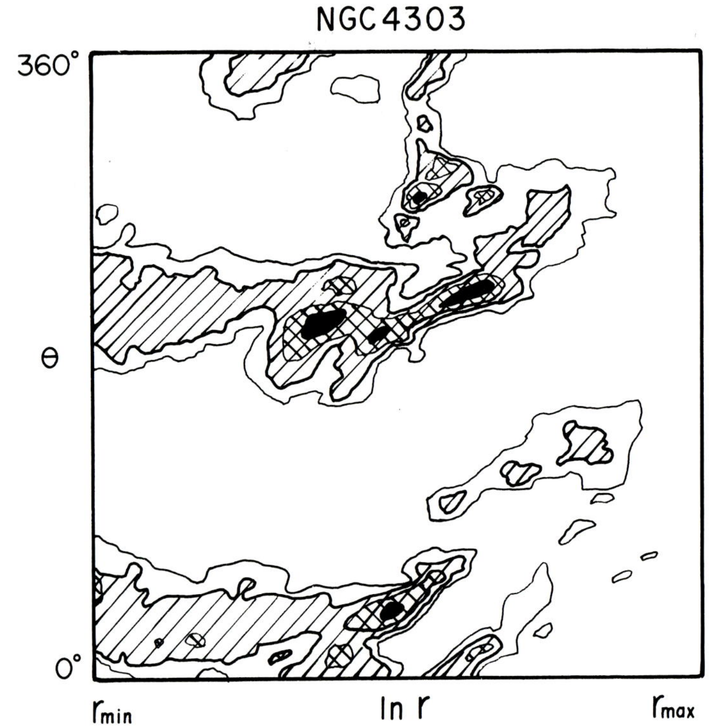


A6.7b: Face-on intensity distribution

A6.7c: Reconstructed face-on image

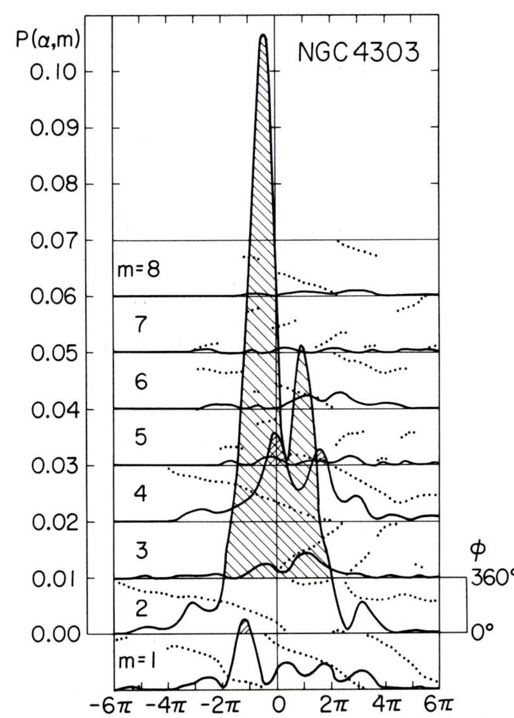


A6.7d: Normalized power spectrum

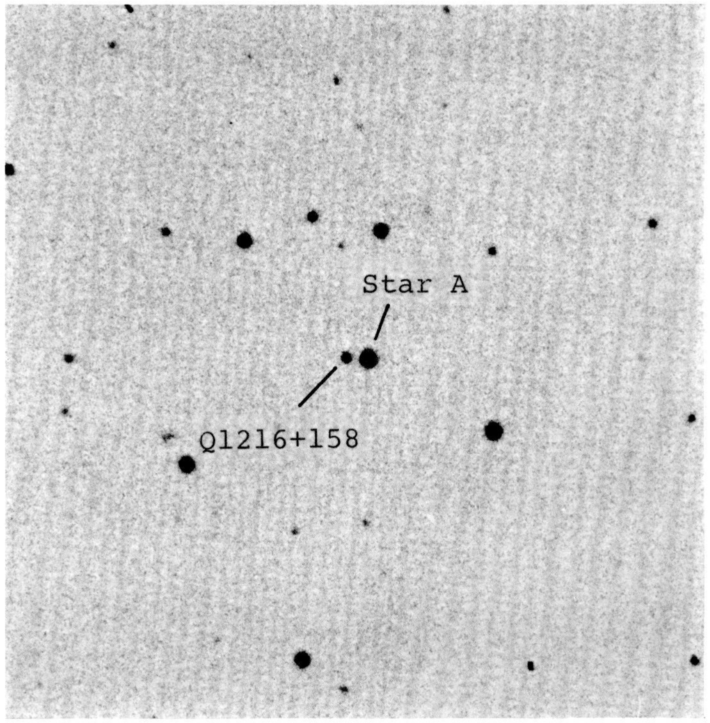


A7.1a: Q1216+158 (QSO) (B)

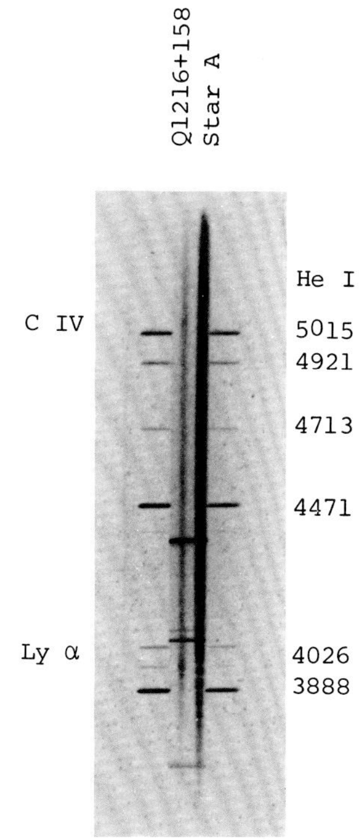


A7.1b: Q1216+158 (Spectrum)

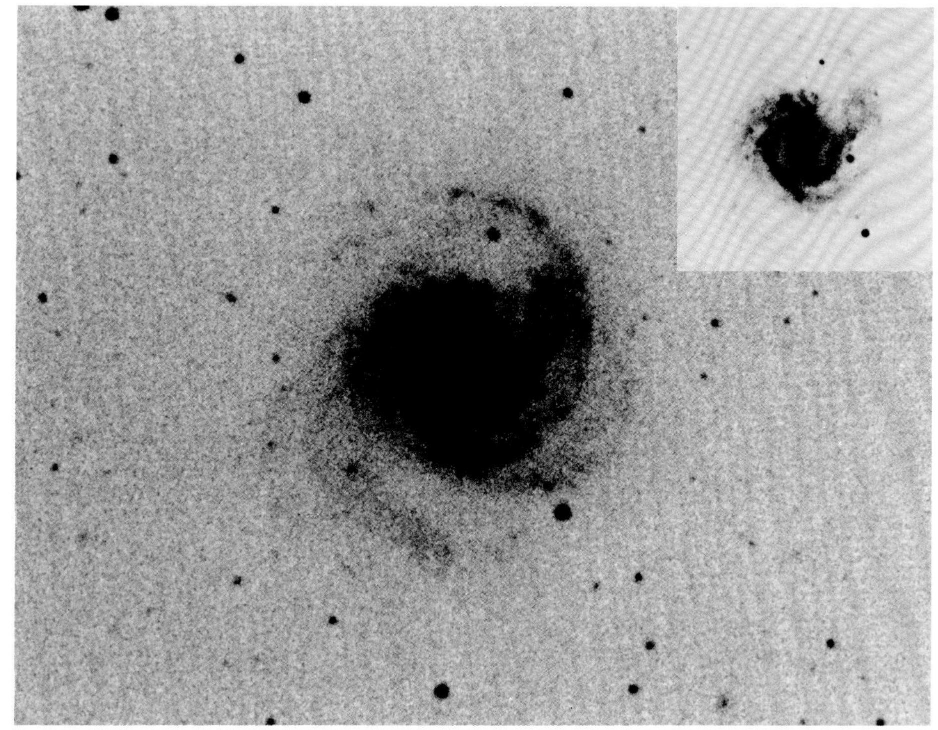

A7.2: NGC 4303 SAB(rs)bc (B)

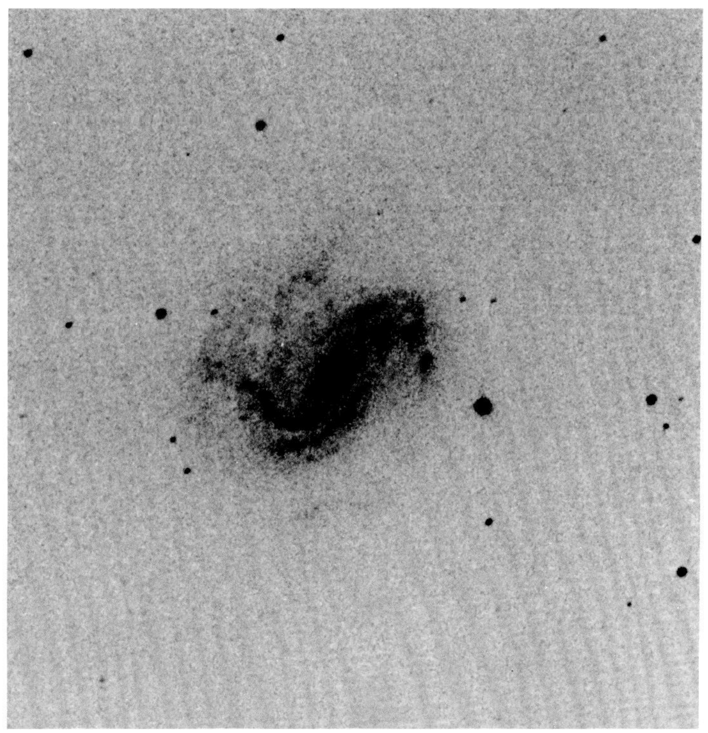

A7.3: NGC 4051 SAB(rs)bc (B)

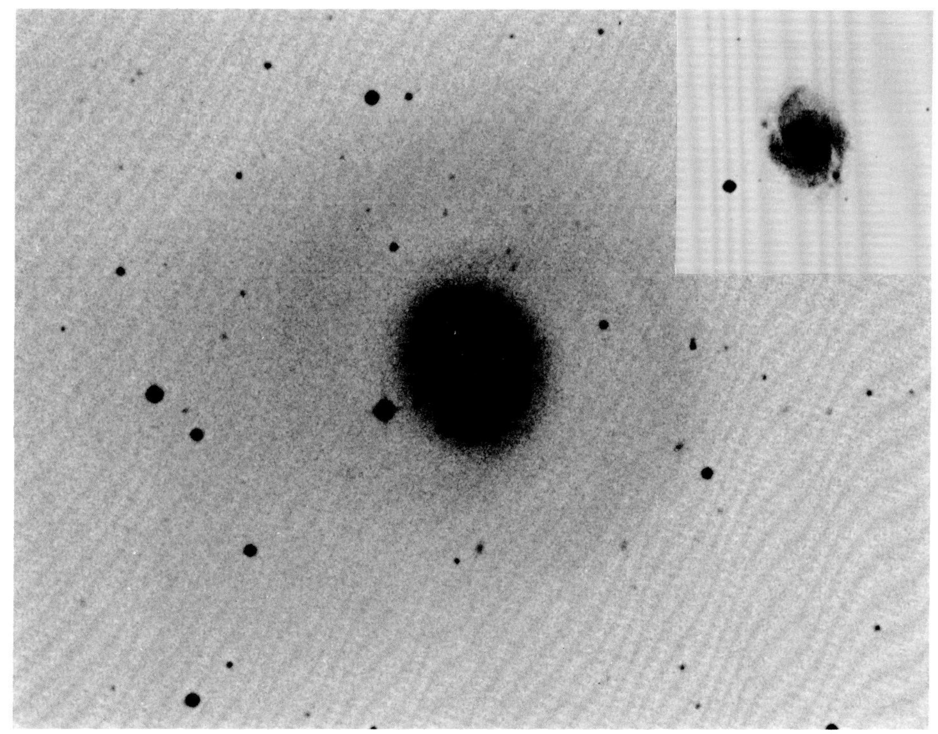

A7.4: NGC 1068 (R)SA(rs)b (V)

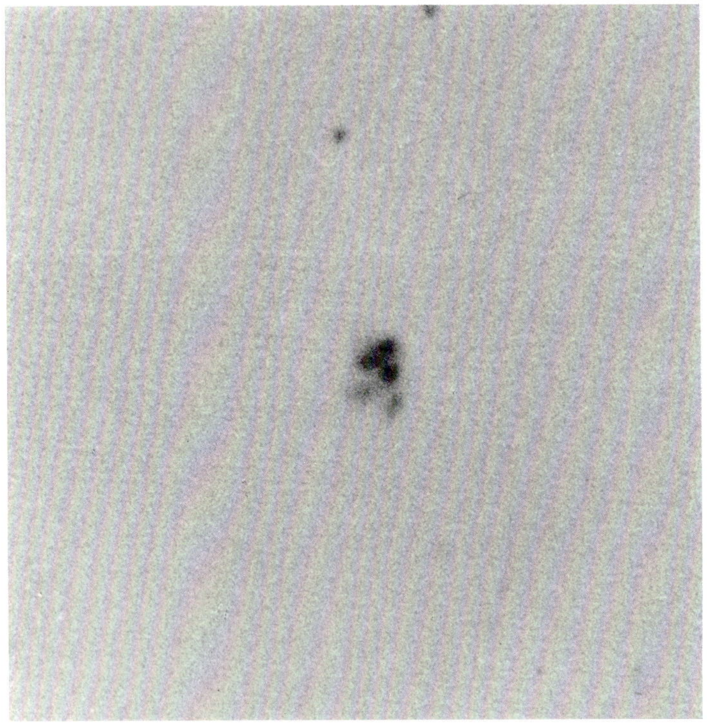

A7.5a: KUG 1626+413 (UGC 10407) (B)

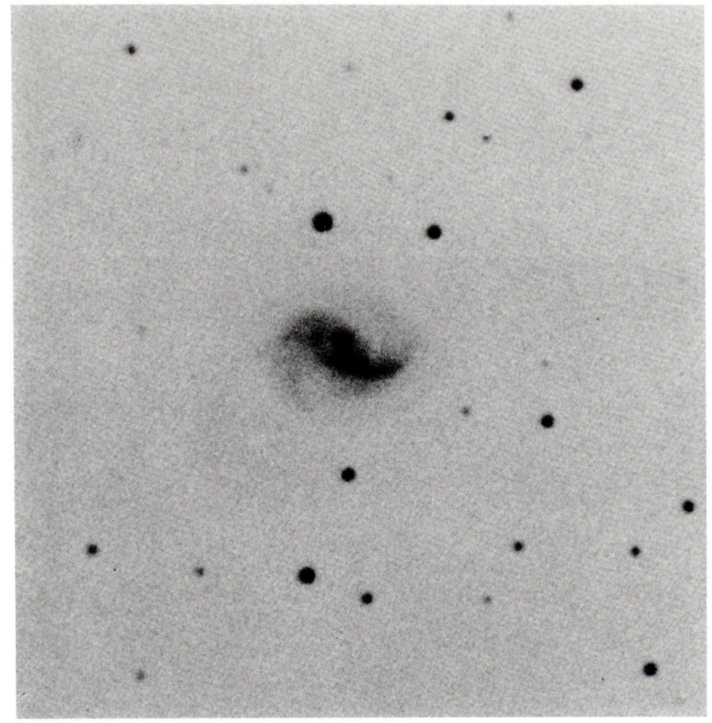

A7.6a: KUG 0239+345 (NGC 1050) (B)

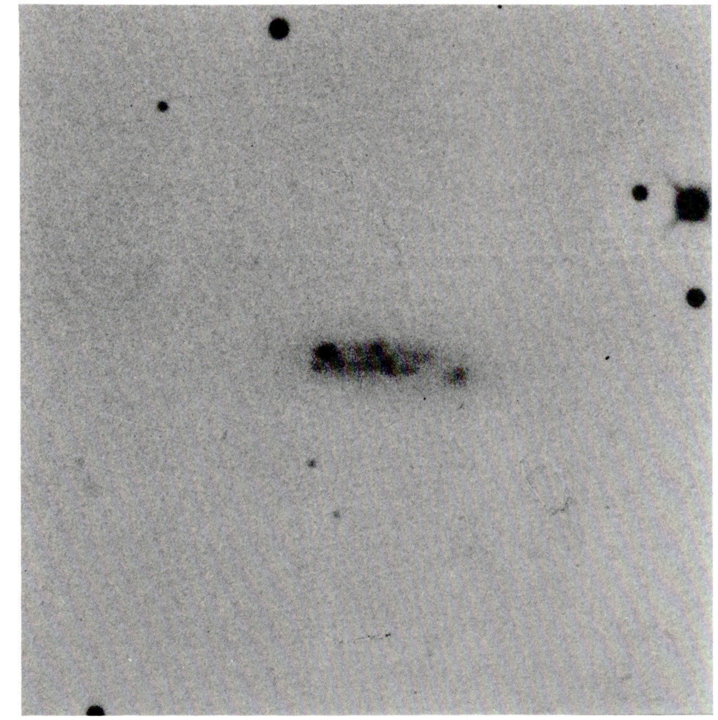

A7.7a: KUG 0225–103 (Mkn 1039) (B)

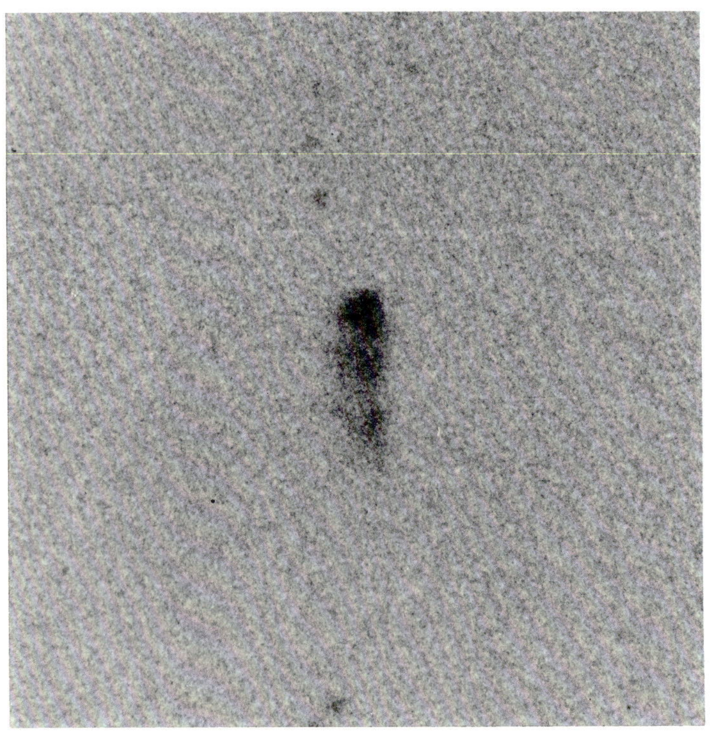

A7.5b: KUG 1626+413 (U,G,R; from top to bottom)

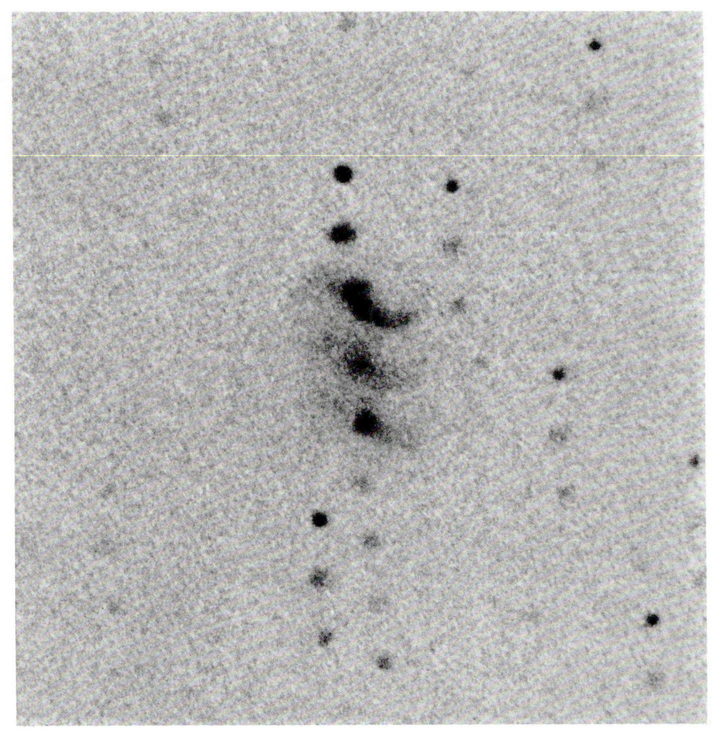

A7.6b: KUG 0239+345 (U,G,R; from top to bottom)

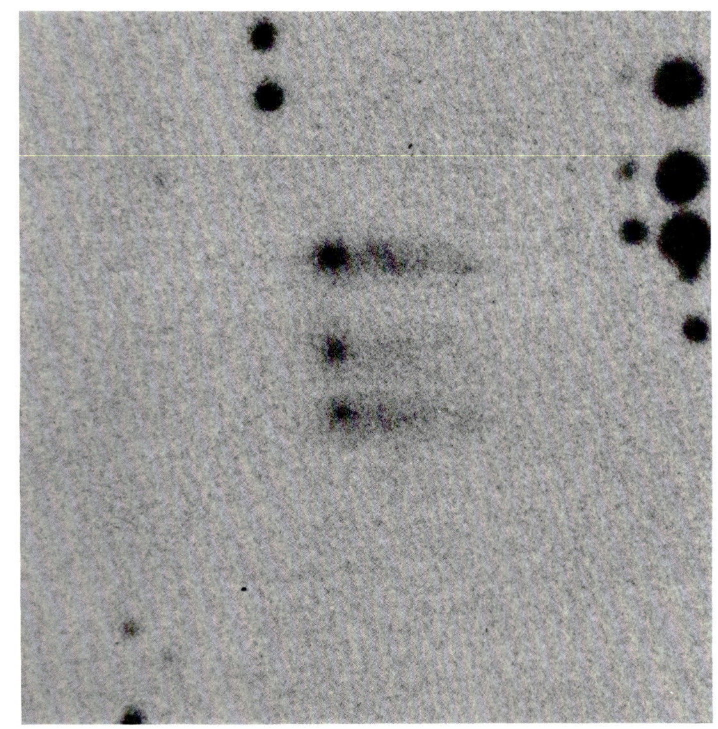

A7.7b: KUG 0225–103 (U,G,R; from top to bottom)

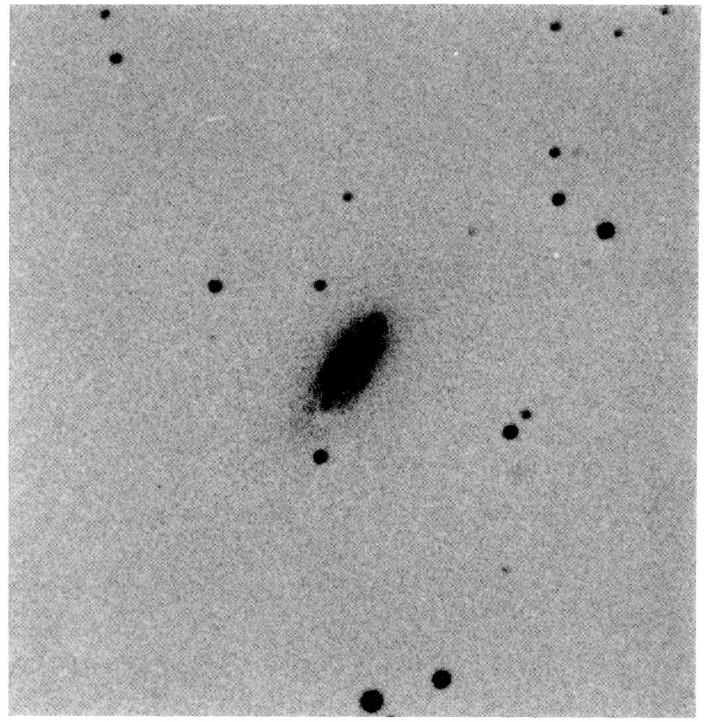

A7.8: KUG 0227+369 (NGC 949) SA(rs)b:? (B)

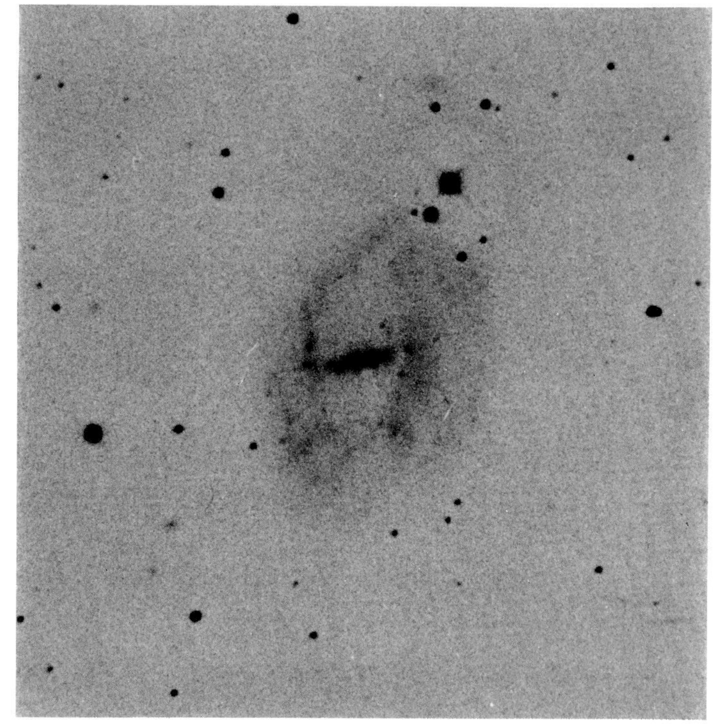

A7.9: KUG 2341+257 (NGC 7741) SB(s)cd (B)

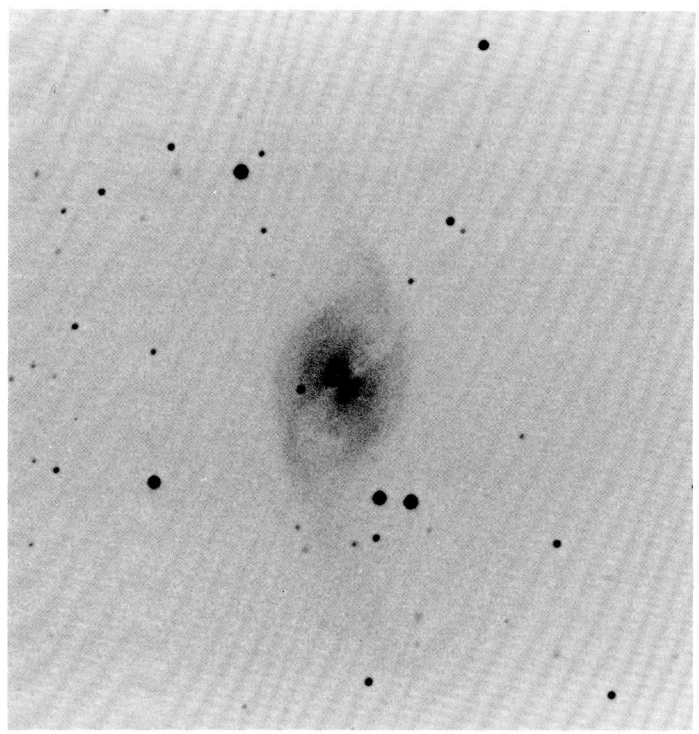

A7.10: NGC 3718 SB(s)aP (V)

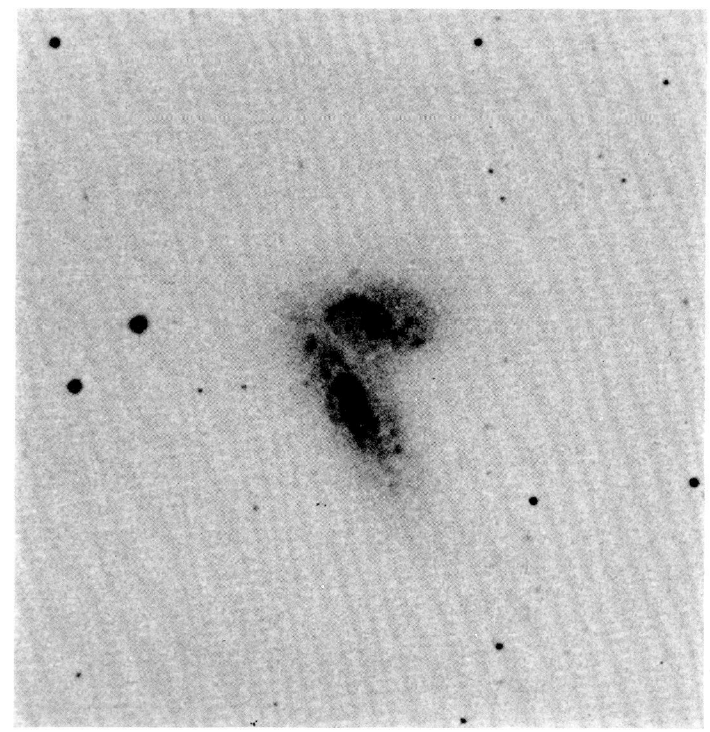

A7.11: NGC 4567 SA(rs)bc,
NGC 4568 SA(rs)bc (B)

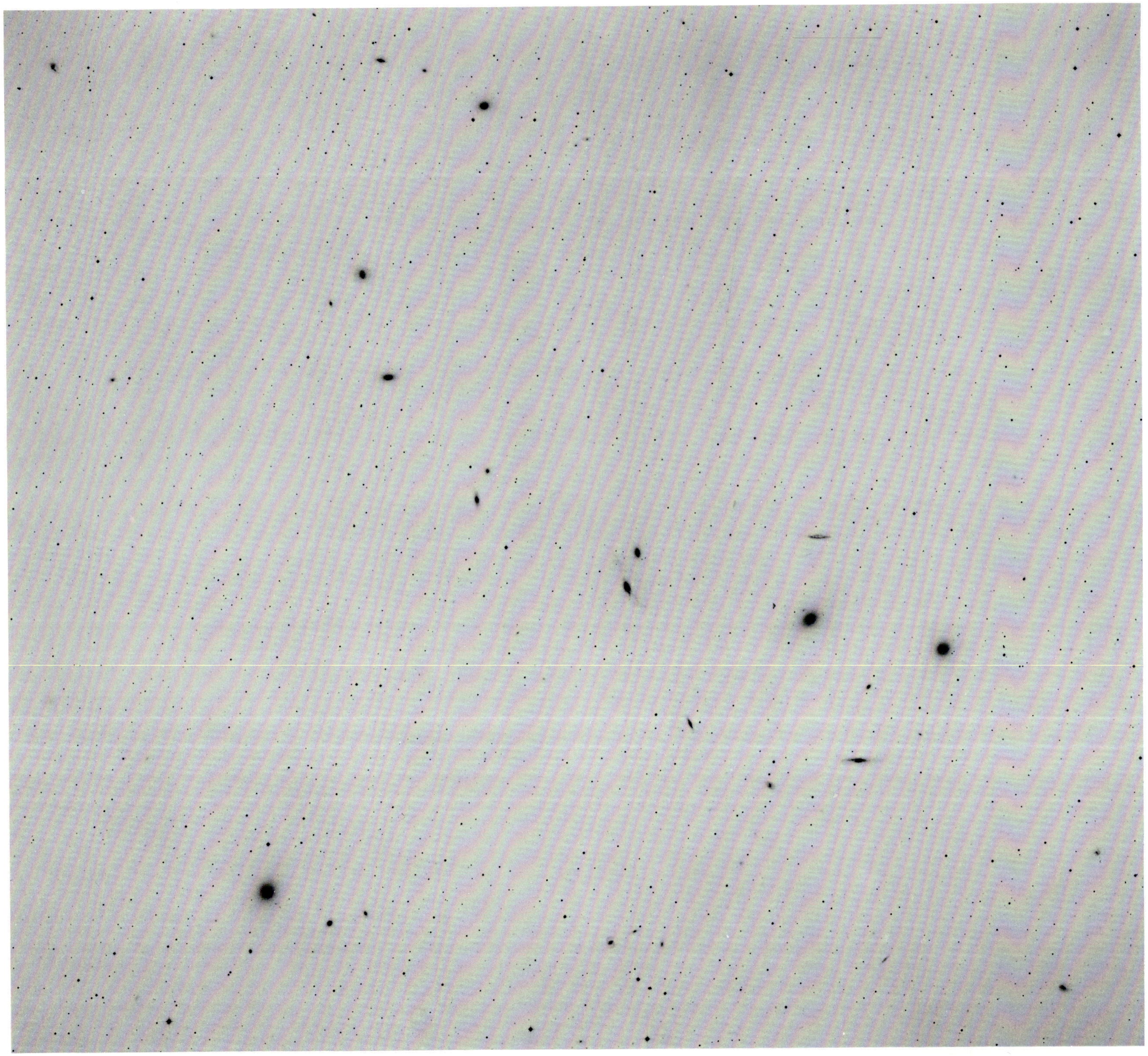

A8.1: Virgo Cluster –, III (V)

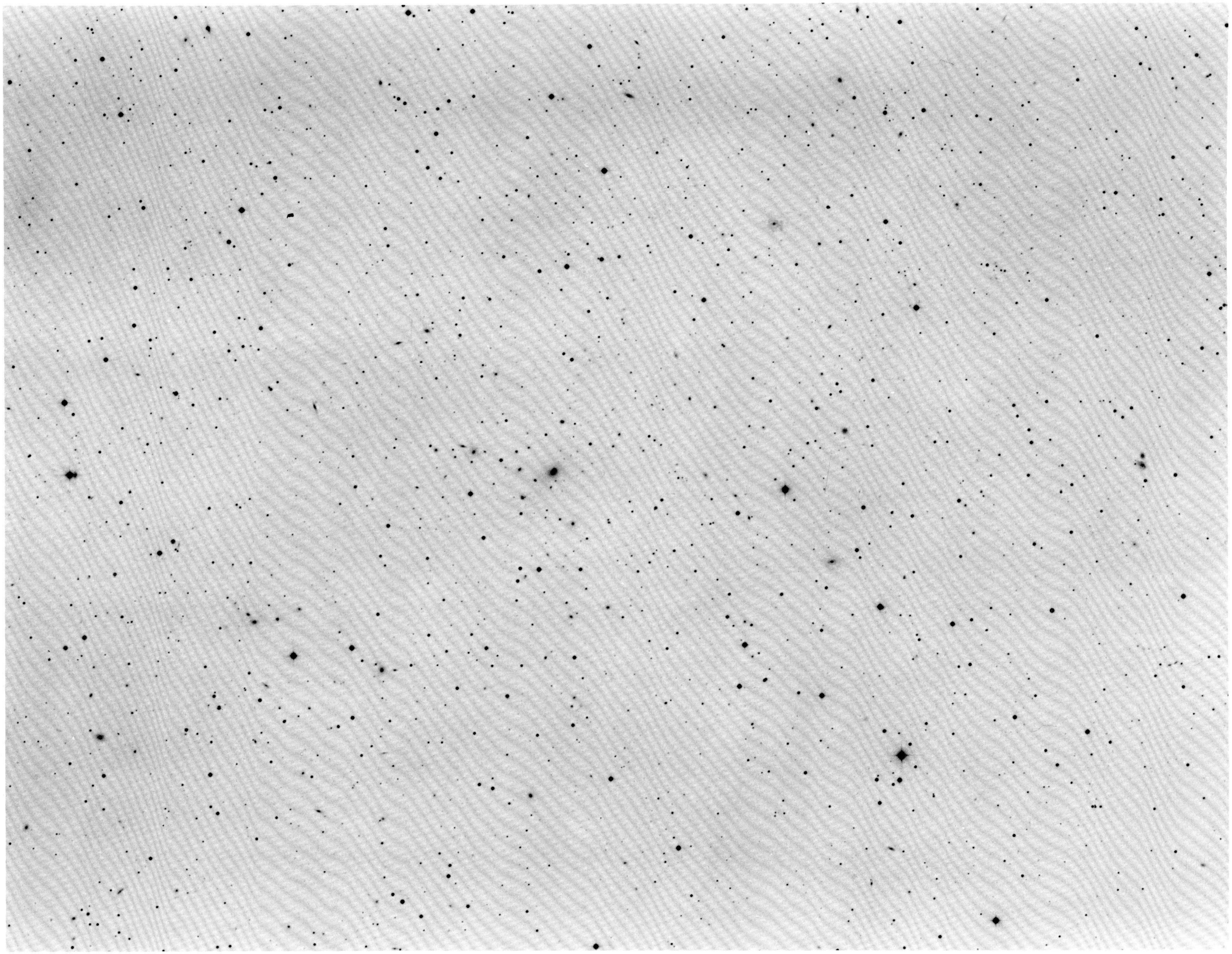

A8.2: A2199 cD, I (B)

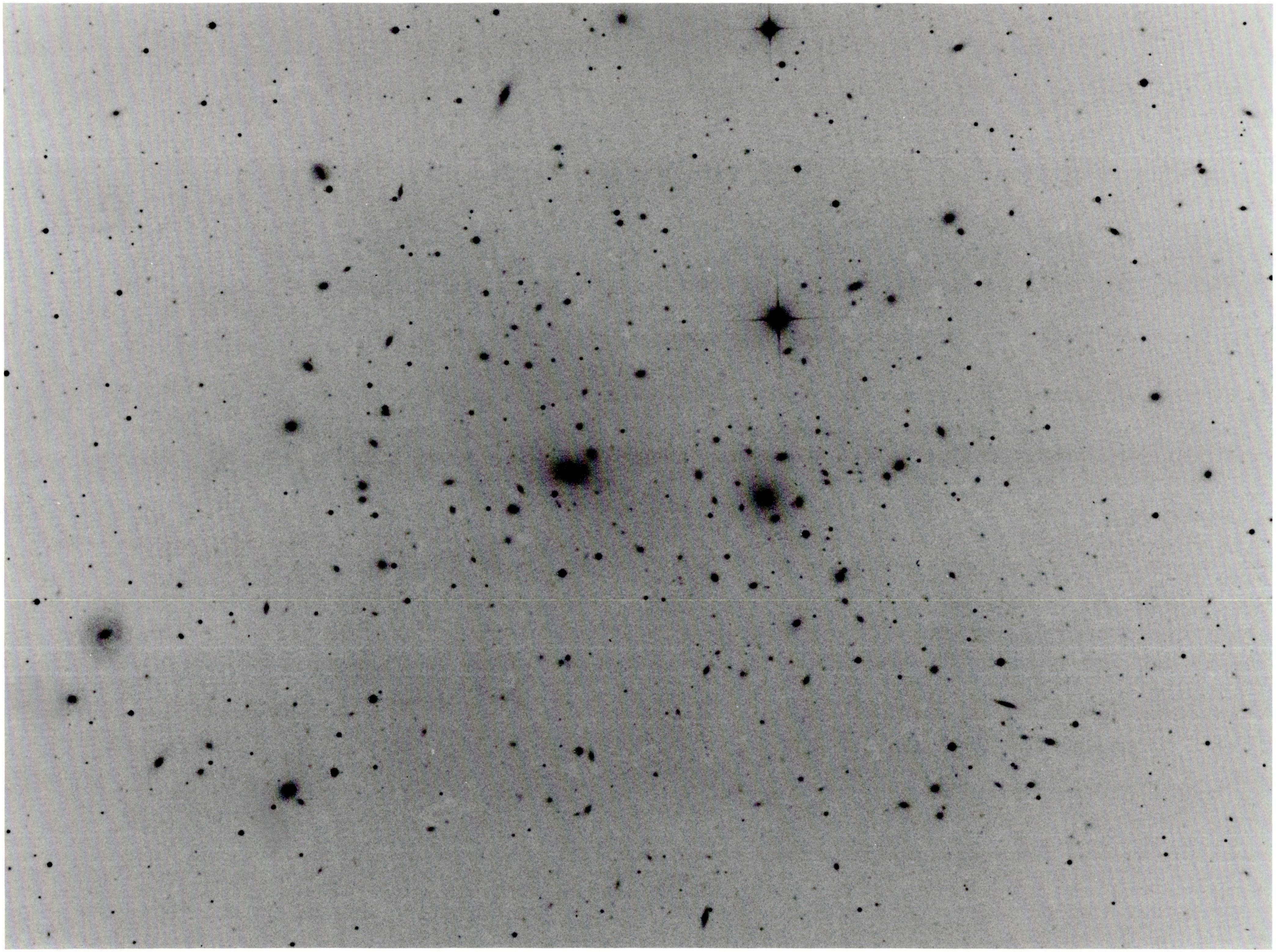

A8.3: A1656 (Coma Cluster) B, II (V)

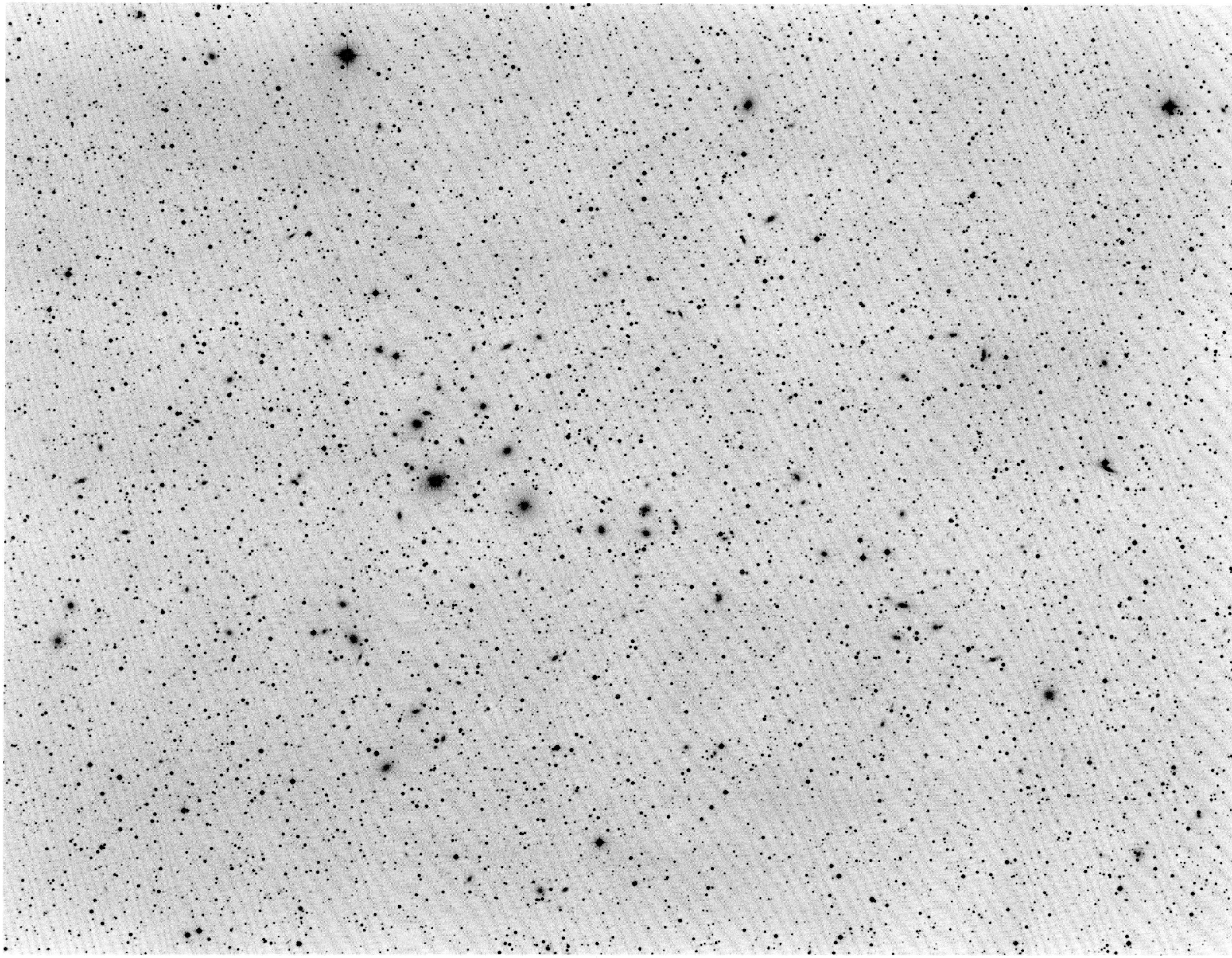

A8.4: A426 (Perseus Cluster) L, III (B)

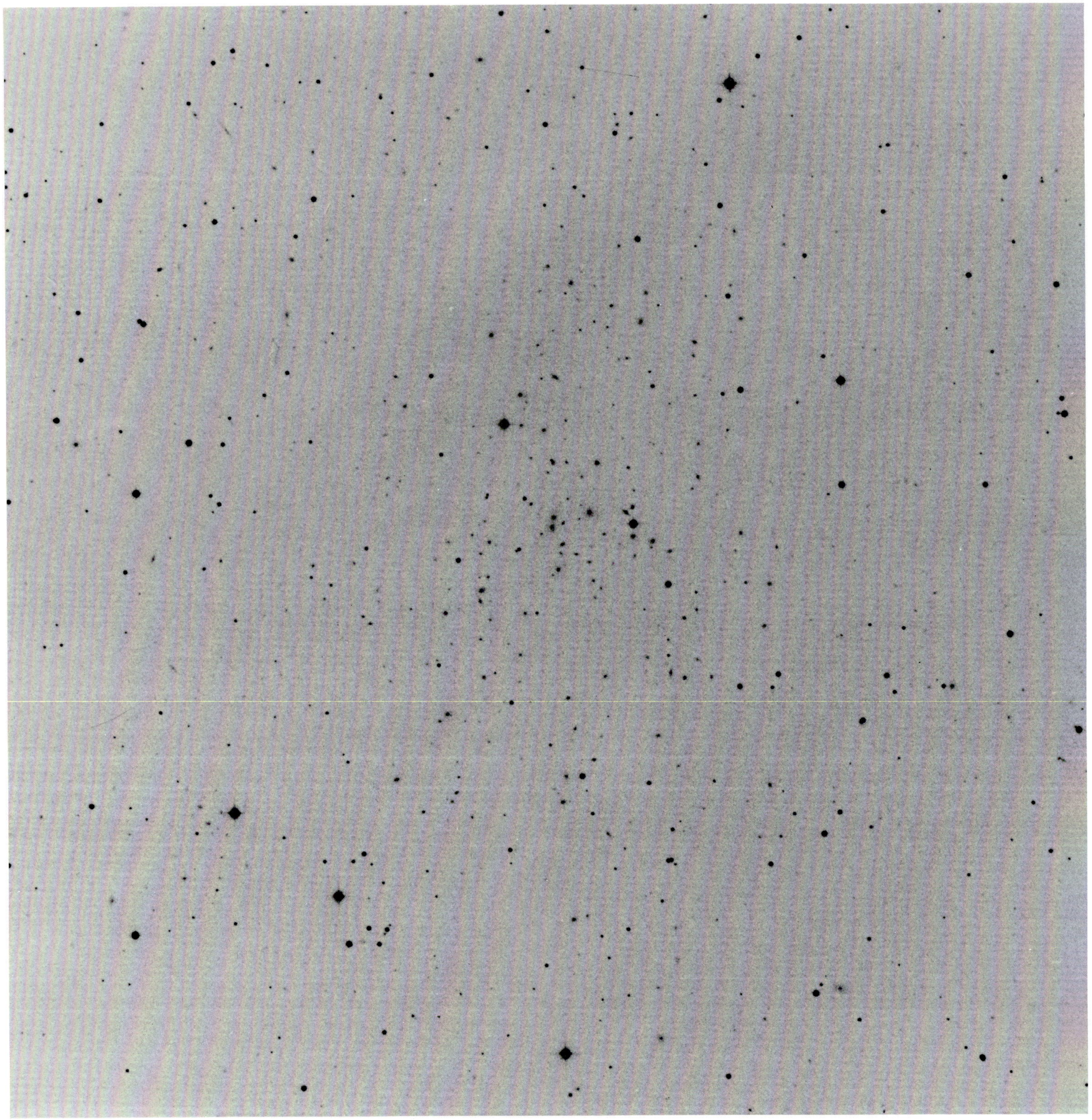

A8.5: A2065 C, III (V)

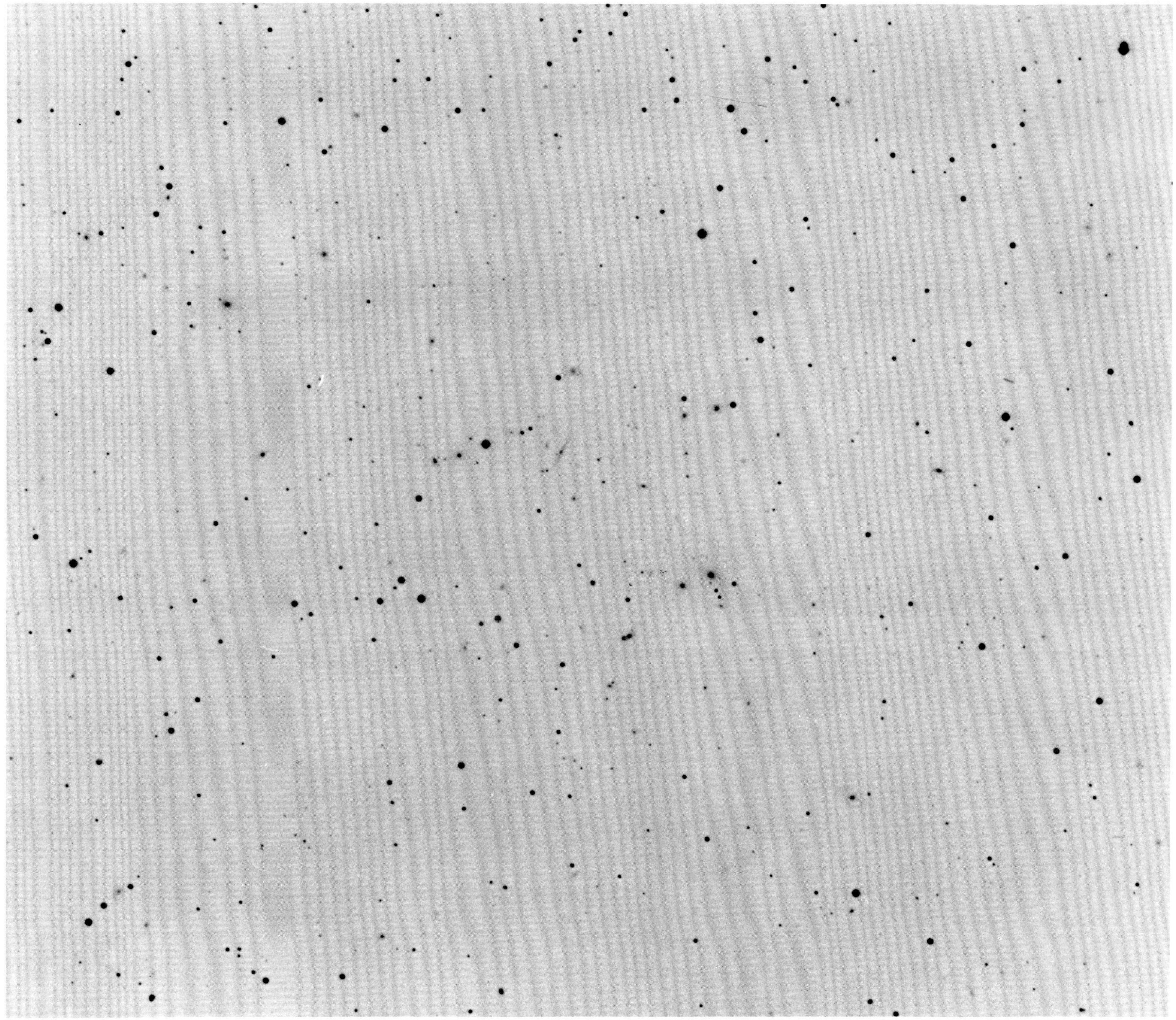

A8.6: A397 F, III (B)

A8.7: A2151 (Hercules Cluster) I, III (J) ⊖

# Part III. Table of Data for the Atlas

# Table 1a. List of Galaxies.

| Name | Rev. Type[1] | DDO Type[2] | Yerkes Type[3] | Fig. No. | Page | Plate No. | Band | Scale of Print (arcsec $mm^{-1}$) | Remarks |
|---|---|---|---|---|---|---|---|---|---|
| NGC 205 | E5P | E6: (t?) | gD4-5 | A1. 3 | 25 | NS-1229 | V | 5.8 | |
| 224 | SA(s)b | Sb I-II | gkS5 | A0. 1 | vi | K-3525 | B | 57.7 | M31 |
| 404 | SA(s)$0^-$: | E0 | Ep* | A1. 4 | 26 | NS-1595 | B | 2.9 | |
| 598 | SA(s)cd | Sc II-III | fS4 | A0. 2a | vii | K-3533 | V | 21.6 | M33 |
| 598 | | | | A0. 2b | vii | K-3526 | U | 21.6 | |
| 891 | SA(s)b?sp | Sb | g?S7 | A4.16 | 45 | K-1003 | B | 11.6 | |
| 949 | SA(rs)b:? | SD? | k?DE6 | A7. 8 | 63 | NS-2587 | B | 3.0 | KUG 0227+369 |
| 1023 | SB(rs)$0^-$ | E7p | kD6p | A1. 5 | 26 | NS-2429 | B | 4.9 | |
| 1050 | ----- | ----- | ----- | A7. 6a | 62 | NS-2715 | B | 2.6 | KUG 0239+345 |
| 1050 | ----- | ----- | ----- | A7. 6b | 62 | K-2524 | UGR | 2.6 | |
| NGC 1068 | (R)SA(rs)b | Sbp | gS2p | A7. 4 | 61 | K-2978 | V | 6.5 | M77, inset: short exposure |
| 2403 | SAB(s)cd | Sc III | fS4 | A3. 6a | 39 | K- 123 | B | 12.1 | |
| 2681 | (R')SAB(rs)0/a | Sa or Snn | kS1* | A1.14 | 28 | NS-1273 | B | 4.8 | |
| 2985 | (R')SA(rs)ab | $Sb^-$ I | gkS3 | A1. 8 | 27 | K-3235 | B | 6.6 | |
| 3031 | SA(s)ab | Sb I-II | gkS4 | A2. 1a | 31 | K-2081 | U | 16.0 | M81 |
| 3031 | | | | A2. 2a | 31 | K-3235 | B | 16.0 | |
| 3031 | | | | A2. 3a | 31 | K-3154 | V | 16.0 | |
| 3031 | | | | A2. 4a | 32 | K-3240 | I | 16.0 | |
| 3034 | I0sp | Pec | E7p-Ip | A1.29 | 30 | NS-1882 | B | 16.0 | M82 |
| 3077 | I0P | E2p | ED3p | A1.28 | 30 | K-3235 | B | 6.6 | |
| NGC 3115 | $S0^-$sp | E6 | kD7* | A4. 3a | 42 | NS-1113 | V | 3.1 | |
| 3338 | SA(s)c | $Sb^+$ II | fS4* | A1.12 | 27 | K-3141 | B | 6.7 | |
| 3351 | SB(r)b | S(B)b II | gB3 | A1.20 | 29 | K-3141 | B | 6.6 | M95 |
| 3379 | E1 | E1 | kED1 | A1. 1 | 25 | K-3141 | B | 6.2 | |
| 3384 | SB(s)$0^-$: | E7 | kD5 | A1. 1 | 25 | K-3141 | B | 6.2 | |
| 3389 | SA(s)c | Sc* III: | | A1. 1 | 25 | K-3141 | B | 6.2 | |
| 3628 | SbPsp | Sbnt | g?S?7 | A4.17 | 45 | K-3242 | B | 7.2 | |
| 3718 | SB(s)aP | Snnt | fgSp | A7.10 | 63 | K-2726 | V | 8.0 | |
| 3780 | SA(s)c: | Sc II | afS2 | A5.27 | 52 | K-2185 | V | 2.6 | |
| 3877 | SA(s)c: | Sbn | afS6 | A5.23 | 51 | K-3539 | V | 5.2 | |
| NGC 3893 | SAB(rs)c: | Scn I: | fgS4 | A1.17 | 28 | K-2713 | B | 6.7 | |
| 3917 | SAcd: | S(n*) | afS6 | A5.31 | 53 | K-3539 | V | 5.2 | |
| 3953 | SB(r)bc | $Sb^+$ I or S(B)$b^+$ I | fgS4 | A3. 4a | 38 | K-2726 | V | 8.3 | |
| 3953 | | | | A5.22 | 51 | K-3539 | V | 7.9 | |
| 3992 | SB(rs)bc | S(B)$b^+$ I | fB4* | A1.21 | 29 | K-2726 | V | 6.7 | |
| 3992 | | | | A5.24 | 51 | K-2185 | V | 10.4 | |
| 4013 | Sb sp | Sb | gk:S7: | A4.14 | 45 | NS-2635 | V | 4.8 | |

| Name | Rev. Type[1] | DDO Type[2] | Yerkes Type[3] | Fig. No. | Page | Plate No. | Band | Scale of Print (arcsec $mm^{-1}$) | Remarks |
|---|---|---|---|---|---|---|---|---|---|
| 4026 | S0sp | E8 | kD7 | A5. 3a | 47 | K-3539 | V | 5.5 | |
| 4051 | SAB(rs)bc | $Sc^{(*)}$ II | fS2 | A7. 3 | 61 | K-3285 | B | 6.7 | |
| 4088 | SAB(rs)bc | Sc* I-II: | a?S4p | A1.16 | 28 | K-2713 | B | 6.6 | |
| NGC 4096 | SAB(rs)c | Sc* II: | afS5 | A1.18 | 28 | K-3285 | B | 6.7 | |
| 4100 | (R')SA(rs)bc | Sbn I-II: | fg:S5p | A5. 5a | 48 | K-3539 | V | 5.4 | |
| 4157 | SAB(s)b?sp | $Sb^{+}$ II: | f?S6 | A5.25 | 52 | K-3539 | V | 7.8 | |
| 4178 | SB(rs)dm | SBc II: | afB5 | A1.24 | 29 | K-3143 | B | 8.3 | |
| 4178 | | | | A5.33 | 53 | K-2186 | V | 5.3 | |
| 4192 | SAB(s)ab | Sbt I-II: | gS6 | A5.26 | 52 | K-3540 | V | 10.4 | M98 |
| 4217 | Sb sp | Sb | gk:S7 | A4.12 | 44 | K-3285 | B | 6.6 | |
| 4217 | | | | A5.29 | 52 | K-3539 | V | 7.8 | |
| 4222 | Sd:sp | ----- | ----- | A4.18 | 45 | K-2729 | B | 6.6 | |
| 4234 | SB(s)dm | $Sb^{+}$ IV | (aS2) | A5. 6a | 48 | K-3547 | V | 2.8 | |
| NGC 4244 | SA(s)cd:sp | $S^{-}$ IV: | af?S?7 | A4. 6a | 43 | K-3764 | B | 8.5 | |
| 4254 | SA(s)c | Sc I | fgS1 | A1.13 | 27 | K-3069 | B | 6.6 | M99 |
| 4261 | E2-3 | E2 | kDE2 | A5.14 | 50 | K-3547 | V | 7.8 | |
| 4267 | $SB(s)0^{-}$? | E2 or Sa | gkD1 | A5. 2a | 46 | K-3540 | V | 5.4 | |
| 4298 | SA(rs)c | S**t | afSD4 | A4.15 | 45 | K-3069 | B | 6.6 | |
| 4302 | Sc:sp | Sc | fS7* | A4.15 | 45 | K-3069 | B | 6.6 | |
| 4303 | SAB(rs)bc | Sc I | fS1 | A6. 7a | 60 | K-3547 | V | 4.7 | M61 |
| 4303 | | | | A7. 2 | 61 | K-3144 | B | 6.7 | inset: short exposure |
| 4307 | Sb sp | $Sb^{-}$II-III | g:SD7 | A4.13 | 45 | K-2186 | V | 5.2 | |
| 4321 | SAB(s)bc | Sc I | fgS1 | A6. 4a | 57 | K-3540 | V | 4.7 | M100 |
| NGC 4340 | $SB(r)0^{+}$ | SBa | gkB4 | A1. 7 | 26 | K-3069 | B | 6.8 | |
| 4340 | | | | A5.12 | 49 | K-3540 | V | 5.2 | |
| 4340 | | | | A6. 5a | 58 | K-3540 | V | 3.3 | |
| 4350 | SA0sp | E7 | gkD7* | A2. 8a | 34 | K-2145 | U | 3.3 | |
| 4350 | | | | A2. 9a | 34 | K-2079 | I | 3.3 | |
| 4350 | | | | A4. 7 | 44 | K-3069 | B | 6.6 | |
| 4350 | | | | A5. 9 | 49 | K-3540 | V | 5.2 | |
| 4365 | E3 | E2 | kE4 | A1. 2 | 25 | K-3143 | B | 6.6 | |
| 4374 | E1 | E1 | kE1 | A3. 1a | 37 | K-3069 | B | 11.0 | M84 |
| 4374 | | | | A5. 1a | 46 | K-3540 | V | 10.8 | |
| NGC 4414 | SA(rs)c? | Scn* II: | fSD4 | A1.11 | 27 | NS-2686 | B | 4.8 | |
| 4417 | SB0:sp | E7 | kD7 | A5. 8 | 49 | K-2186 | V | 5.2 | |
| 4419 | SB(s)a sp | Ep | gkSD6 | A5.17 | 50 | K-3540 | V | 5.3 | |

| Name | Rev. Type[1)] | DDO Type[2)] | Yerkes Type[3)] | Fig. No. | Page | Plate No. | Band | Scale of Print (arcsec mm$^{-1}$) | Remarks |
|---|---|---|---|---|---|---|---|---|---|
| 4429 | SA(r)0$^+$ | ? | kD5 | A1. 6 | 26 | K-3143 | B | 6.6 | inset: short exposure |
| 4429 | | | | A3. 2a | 37 | K-3143 | B | 8.2 | |
| 4440 | SB(rs)a | ----- | ----- | A5.13 | 50 | K-3540 | V | 2.6 | |
| 4442 | SB(s)0$^0$ | E5p | kD6 | A5. 7 | 49 | K-2186 | V | 5.2 | |
| 4450 | SA(s)ab | Sbn | gS3 | A5.18 | 50 | K-3540 | V | 7.8 | |
| 4469 | SB(s)0/a?sp | Sp | gD7 | A4.10 | 44 | K-3143 | B | 6.6 | |
| 4469 | | | | A5.20 | 51 | K-2186 | V | 5.2 | |
| NGC 4472 | E2 | E4 | kE2 | A2. 6a | 33 | K-2230 | U | 12.4 | M49 |
| 4472 | | | | A2. 7a | 33 | K-2087 | I | 12.4 | |
| 4472 | | | | A6. 6a | 59 | K-2186 | V | 4.8 | |
| 4477 | SB(s)0:? | S(B)a | kBD1 | A5.15 | 50 | K-3540 | V | 5.2 | |
| 4483 | SB(s)0/a: | S(B)a? | ----- | A5.11 | 49 | K-2186 | V | 2.6 | |
| 4485 | IB(s)mP | Ir III-IV: | a?I | A1.26 | 30 | NS-2006 | B | 4.8 | |
| 4486 | E$^+$0-1P | E1 | kE1 | A5.10 | 49 | K-3540 | V | 13.0 | M87 |
| 4496A | SB(rs)m | S(B)c III: | aBS1 | A1.25 | 29 | K-3144 | B | 6.7 | |
| 4496A | | | | A5.30 | 52 | K-3547 | V | 5.3 | |
| 4496B | IB(s)m: | ----- | afI | A1.25 | 29 | K-3144 | B | 6.7 | |
| NGC 4501 | SA(rs)b | Sb$^+$ I | gS4 | A2.10a | 34 | K-2145 | U | 8.3 | M88 |
| 4501 | | | | A2.11a | 35 | K-2079 | I | 8.3 | |
| 4501 | | | | A6. 2a | 55 | K-3540 | V | 4.7 | |
| 4519 | SB(rs)d | Sc III | afS1 | A1.23 | 29 | K-2222 | B | 6.6 | |
| 4522 | SB(s)cd:sp | Scp | af?S6? | A5.32 | 53 | K-2220 | V | 2.6 | |
| 4527 | SAB(s)bc | Sb$^+$(n) II | gS6 | A5.21 | 51 | K-3740 | V | 7.8 | |
| 4535 | SAB(s)c | S(B)c I: | afS2 | A2.12a | 35 | K-2221 | U | 7.1 | |
| 4535 | | | | A2.13a | 35 | K-3244 | I | 7.1 | |
| 4535 | | | | A3. 5a | 39 | K-2222 | B | 5.6 | |
| 4535 | SAB(s)c | S(B)c I: | afS2 | A6. 1a | 54 | K-2220 | V | 4.8 | |
| NGC 4548 | SB(rs)b | SBbn | gB(3) | A5.19 | 51 | K-2730 | V | 7.9 | |
| 4565 | SA(s)b?sp | Sb I: | gkS7 | A4. 5a | 43 | K-1033 | B | 8.5 | |
| 4567 | SA(rs)bc | Sc**t | afS3 | A7.11 | 63 | K-2222 | B | 6.6 | |
| 4568 | SA(rs)bc | Sc**t | afS5* | A7.11 | 63 | K-2222 | B | 6.6 | |
| 4569 | SAB(rs)ab | Sb$^+$n | aS5 | A1.15 | 28 | K-3151 | B | 7.6 | M90 |
| 4569 | | | | A5. 4a | 47 | K-2730 | V | 10.8 | |
| 4571 | SA(r)d | Snn* | fSD2 | A5.28 | 52 | K-2730 | V | 5.3 | |
| 4579 | SAB(rs)b | Sbn | kDS1 | A6. 3a | 56 | K-2220 | V | 4.8 | M58 |
| 4594 | SA(s)a sp | Sb$^-$ | kDS6 | A4. 4a | 42 | NS-2448 | V | 5.3 | M104 |
| 4623 | SB0$^+$?sp | E:5 | ----- | A5.16 | 50 | K-2220 | V | 2.7 | |

| Name | Rev. Type[1] | DDO Type[2] | Yerkes Type[3] | Fig. No. | Page | Plate No. | Band | Scale of Print (arcsec $mm^{-1}$) | Remarks |
|---|---|---|---|---|---|---|---|---|---|
| NGC 4654 | SAB(rs)cd | Sc$^{(*)}$ II | aS3 | A1.19 | 28 | K-2731 | B | 6.6 | |
| 4689 | SA(rs)bc | Sb$^+$ II: | fSD2 | A1.10 | 27 | K-3151 | B | 6.6 | |
| 4698 | SA(s)ab | Sb$^-$ II: | kS5 | A3. 3a | 38 | K-2222 | B | 5.6 | |
| 4710 | SA(r)$0^+$?sp | E8 | k?D?7 | A4. 9 | 44 | K-3151 | B | 6.6 | inset: short exposure |
| 4736 | (R)SA(r)ab | Sb$^-$p II: | gDS3 | A1. 9 | 27 | NS- 791 | B | 4.6 | M94 |
| 4762 | SB(r)$0^0$?sp | Snnt: | kD7 | A4. 2a | 41 | K-2222 | B | 4.9 | |
| 5194 | SA(s)bcP | Sc(t) I | fgS1 | A0. 3a | viii | NS-2007 | B | 4.5 | M51 |
| 5194 | | | | A0. 3b | viii | NS-2356 | I | 4.6 | M51 |
| 5195 | I0P | Pec(t) | f?Ep? | A0. 3a | viii | NS-2007 | B | 4.5 | M51 |
| NGC 5195 | | | | A0. 3b | viii | NS-2356 | I | 4.6 | M51 |
| NGC 5457 | SAB(rs)cd | Sc I | fS1 | A2.14a | 36 | K-2777 | U | 2.0 | M101 |
| 5457 | | | | A2.14b | 36 | K-3162 | B | 2.0 | |
| 5457 | | | | A2.14c | 36 | K-3269 | I | 2.0 | |
| 5866 | SA$0^+$sp | E6p | kD7p | A4. 1a | 40 | NS-1886 | B | 2.4 | M102? |
| 7332 | S0Psp | E7 | k?D7* | A4. 8 | 44 | NS-2068 | V | 2.4 | |
| 7479 | SB(s)c | SBb$^+$ I | fB2* | A1.22 | 29 | NS-1603 | B | 4.8 | |
| 7741 | SB(s)cd | SBc II | afB* | A7. 9 | 63 | NS-2413 | B | 4.8 | KUG2341+257 |
| 7814 | SA(s)ab:sp | Sb$^-$ | kS7* or kD7p* | A4.11 | 44 | K-2487 | B | 6.6 | |
| IC 3059 | Im | ----- | ----- | A5.35 | 53 | K-3540 | V | 2.6 | |
| 3322A | SB(s)cd:sp | ----- | ----- | A5.36 | 53 | K-3547 | V | 2.6 | |
| IC 3476 | IB(s)m: | ----- | ----- | A1.27 | 30 | K-3540 | V | 2.7 | |
| 3583 | IBm | ----- | ----- | A1.15 | 28 | K-3151 | B | 7.6 | |
| A1155+51 | SB(s)m | ----- | ----- | A5.34 | 53 | K-3539 | V | 2.7 | |
| Mkn 1039 | ----- | ----- | ----- | A7. 7a | 62 | NS-2723 | B | 2.7 | KUG0225 - 103 |
| Mkn 1039 | ----- | ----- | ----- | A7. 7b | 62 | K-2929 | UGR | 2.7 | |
| UGC 10407 | ----- | ----- | ----- | A7. 5a | 62 | NS-2552 | B | 2.8 | KUG1626+413 |
| 10407 | ----- | ----- | ----- | A7. 5b | 62 | K-2168 | UGR | 2.8 | |
| Q1216+158 | (QSO) | ----- | ----- | A7. 1a | 61 | K-3069 | B | 6.3 | |

1) Revised Hubble type given in RC2 catalogue (C5)
2) Type given by van den Bergh (G61)
3) Type given by Morgan (G39, or G38 when indicated by an asterisk)

# Table 1b. List of Clusters of Galaxies.

| No.[1] | Revised R-S type[2] | B-M type[3] | Fig. No. | Page | Plate | Band | Scale of Print (arcsec $mm^{-1}$) | Remarks |
|---|---|---|---|---|---|---|---|---|
| A 397 | F | III | A8.6 | 69 | K-3399 | B | 9.9 | ($2^h54^m$, +15°45′)[4] |
| A 426 | L | III | A8.4 | 67 | K-3535 | B | 15.2 | Perseus Cluster |
| A 1656 | B | II | A8.3 | 66 | K-3683 | V | 9.9 | Coma Cluster |
| A 2065 | C | III | A8.5 | 68 | K-3774 | V | 9.6 | ($15^h21^m$, +27°50′)[4] |
| A 2151 | I | III | A8.7 | 70 | K-1520 | J | 14.3 | Hercules Cluster |
| A 2199 | cD | I | A8.2 | 65 | K-3688 | B | 15.7 | ($16^h27^m$, +39°40′)[4] |
| ---- | --- | III | A8.1 | 64 | K-2186 | V | 36.9 | Virgo Cluster |

1) Number given by Abell (G1)
2) Type given by Struble and Rood (G53)
3) Type given by Bautz and Morgan (G7)
4) Approximate position of the cluster center

# Table 2a. Plate List of the Kiso Schmidt Telescope.

| Plate No. | Emulsion | Filter | Exposure (min) | Date |
|---|---|---|---|---|
| K- 123 | IIa-O | GG385 | 60 | 1975, Dec. 3 |
| K-1003 | 103a-O | GG385 | 60 | 1976, Dec. 17 |
| K-1033 | 103a-O | GG385 | 20 | 1977, Jan. 23 |
| K-1112 | IIa-O | GG385 | 5 | 1977, May 16 |
| K-1113 | IIa-O | GG385 | 40 | 1977, May 16 |
| K-1520 | IIIa-J$^+$ | GG385 | 213 | 1978, Mar. 17 |
| K-2079 | I-N$^+$ | RG695 | 60 | 1979, Feb. 28 |
| K-2081 | IIa-O$^+$ | UG1 | 90 | 1979, Mar. 1 |
| K-2087 | I-N$^+$ | RG695 | 60 | 1979, Mar. 2 |
| K-2090 | I-N$^+$ | RG695 | 60 | 1979, Mar. 22 |
| K-2145 | IIa-O | UG1 | 70 | 1979, Apr. 3 |
| K-2168 | 103a-E$^+$ | MUL3* | 155 | 1979, Apr. 20 |
| K-2185 | IIa-D | G495 | 30 | 1979, May 2 |
| K-2186 | IIa-D | GG495 | 40 | 1979, May 2 |
| K-2220 | IIa-D | GG495 | 30 | 1979, May 23 |
| K-2221 | IIa-O | UG1 | 60 | 1979, May 23 |
| K-2222 | IIa-O | GG385 | 30 | 1979, May 23 |
| K-2230 | IIa-O | UG1 | 60 | 1979, May 27 |
| K-2231 | IIa-O | GG385 | 30 | 1979, May 27 |
| K-2487 | IIa-O | GG385 | 30 | 1979, Dec. 14 |
| K-2524 | 103a-E$^+$ | MUL3* | 160 | 1979, Dec. 20 |
| K-2713 | IIa-O | GG385 | 25 | 1980, Mar. 13 |
| K-2726 | IIa-D | GG495 | 30 | 1980, Mar. 16 |
| K-2729 | IIa-O$^+$ | GG385 | 25 | 1980, Mar. 16 |
| K-2730 | IIa-D | GG495 | 30 | 1980, Mar. 16 |
| K-2731 | IIa-O | GG385 | 25 | 1980, Mar. 16 |
| K-2777 | IIa-O$^+$ | UG1 | 65 | 1980, Apr. 15 |
| K-2929 | 103a-E$^+$ | MUL3* | 104 | 1980, Oct. 31 |
| K-2978 | IIa-D | GG495 | 50 | 1980, Nov. 10 |
| K-3069 | IIa-O | GG385 | 35 | 1980, Dec. 17 |
| K-3141 | IIa-O$^+$ | GG385 | 40 | 1981, Feb. 5 |
| K-3143 | IIa-O$^+$ | GG385 | 35 | 1981, Feb. 9 |
| K-3144 | IIa-O$^+$ | GG385 | 40 | 1981, Feb. 5 |
| K-3151 | IIa-O | GG385 | 35 | 1981, Feb. 6 |
| K-3154 | IIa-D$^+$ | GG495 | 30 | 1981, Feb. 7 |
| K-3162 | IIa-O$^+$ | GG385 | 25 | 1981, Feb. 9 |
| K-3234 | I-N$^+$ | RG695 | 60 | 1981, Apr. 6 |
| K-3235 | IIa-O$^+$ | GG385 | 30 | 1981, Apr. 6 |
| K-3240 | I-N$^+$ | RG695 | 75 | 1981, Apr. 7 |
| K-3242 | IIa-O | GG385 | 25 | 1981, Apr. 7 |
| K-3244 | I-N$^+$ | RG695 | 70 | 1981, Apr. 7 |
| K-3269 | IV-N$^+$ | RG695 | 60 | 1981, Apr. 28 |
| K-3285 | IIa-O$^+$ | GG385 | 20 | 1981, May 1 |
| K-3399 | IIa-O | GG385 | 35 | 1981, Oct. 27 |
| K-3525 | IIa-O | GG385 | 35 | 1981, Dec. 25 |
| K-3526 | IIa-O$^+$ | UG1 | 70 | 1981, Dec. 25 |
| K-3533 | IIa-D | GG495 | 40 | 1981, Dec. 26 |
| K-3535 | IIa-O | GG385 | 35 | 1981, Dec. 26 |
| K-3539 | IIa-D | GG495 | 45 | 1981, Dec. 26 |
| K-3540 | IIa-D | GG495 | 45 | 1981, Dec. 26 |
| K-3547 | IIa-D | GG495 | 43 | 1981, Dec. 27 |
| K-3683 | IIa-D | GG495 | 50 | 1982, Mar. 22 |
| K-3688 | IIa-O | GG385 | 30 | 1982, Mar. 22 |
| K-3740 | IIa-D | GG495 | 40 | 1982, Apr. 23 |
| K-3764 | IIa-O$^+$ | GG385 | 30 | 1982, May 23 |
| K-3774 | IIa-D$^+$ | GG495 | 30 | 1982, May 24 |

*MUL3 = UG1, 50BP and RG610

## Table 2b. Plate List of the Okayama 188 cm Reflector.

| Plate No. | Emulsion | Filter | Exposure (min) | Date |
|---|---|---|---|---|
| NS- 791 | 103a-O | L39 | 30 | 1967, Apr. 5 |
| NS- 917 | 103a-D | Y50 | 40 | 1968, Mar. 3 |
| NS-1113 | 103a-D | Y50 | 33 | 1969, Mar. 17 |
| NS-1229 | 103a-D | Y50 | 40 | 1969, Oct. 16 |
| NS-1273 | 103a-O | L39 | 22 | 1970, Apr. 4 |
| NS-1403 | 103a-O | L39 | 30 | 1971, Mar. 27 |
| NS-1595 | 103a-O | L39 | 40 | 1972, Oct. 4 |
| NS-1603 | 103a-O | L39 | 40 | 1972, Oct. 6 |
| NS-1847 | 103a-D | Y50 | 45 | 1974, Mar. 19 |
| NS-1872 | 103a-O | L39 | 40 | 1974, Mar. 23 |
| NS-1882 | 103a-O | L39 | 30 | 1974, Apr. 17 |
| NS-1886 | 103a-O | L39 | 35 | 1974, Apr. 17 |
| NS-2006 | 103a-O | L39 | 30 | 1975, Apr. 8 |
| NS-2007 | 103a-O | L39 | 30 | 1975, Apr. 8 |
| NS-2068 | 103a-D | Y50 | 25 | 1975, Nov. 3 |
| NS-2132 | 103a-O | L39 | 10 | 1976, Mar. 28 |
| NS-2356 | I-N$^+$ | R68 | 150 | 1978, Apr. 4 |
| NS-2364 | 103a-O | L39 | 40 | 1978, Apr. 7 |
| NS-2371 | 103a-O | L39 | 52 | 1978, Apr. 8 |
| NS-2413 | 103a-O | L39 | 40 | 1978, Oct. 1 |
| NS-2429 | 103a-O | L39 | 30 | 1978, Oct. 3 |
| NS-2447 | 103a-O | L39 | 10 | 1979, Jan. 30 |
| NS-2448 | 103a-D | Y50 | 15 | 1979, Jan. 30 |
| NS-2539 | IIa-O | L39 | 40 | 1980, Mar. 19 |
| NS-2542 | IIa-O | L39 | 3 | 1980, Mar. 19 |
| NS-2552 | IIa-O | L39 | 30 | 1980, Apr. 15 |
| NS-2587 | IIa-O | L39 | 50 | 1980, Oct. 9 |
| NS-2635 | 103a-D | Y50 | 38 | 1981, Feb. 5 |
| NS-2686 | 103a-O | L39 | 40 | 1981, May 1 |
| NS-2715 | IIa-O$^+$ | L39 | 40 | 1981, Oct. 26 |
| NS-2723 | IIa-O$^+$ | L39 | 30 | 1981, Oct. 27 |
| IS- 963 | IIa-O$^+$ | NONE | 120 | 1982, Mar. 21 |

## Table 3. Data for Isophotal Maps.

| Galaxy | Fig. No. | Plate | Faintest isophote[1] | Isophote intervals[2] |
|---|---|---|---|---|
| NGC 3031 | A2. 1b | K-2081 | 0.0398* | 1.5, 1.0, 0.5×7 |
| NGC 3031 | A2. 2b | K-3235 | 0.0398* | 1.5, 1.0, 0.5×7 |
| NGC 3031 | A2. 3b | K-3154 | 0.0398* | 1.5, 1.0, 0.5×7 |
| NGC 3031 | A2. 4b | K-3240+3243 | 0.0398* | 1.5, 1.0, 0.5×7 |
| NGC 4472 | A2. 6b | K-2230 | 0.0398* | 1.5, 1.0, 0.5×7 |
| NGC 4472 | A2. 7b | K-2087 | 0.0398* | 1.5, 1.0, 0.5×7 |
| NGC 4350 | A2. 8b | K-2145 | 0.0398* | 1.5, 1.0, 0.5×7 |
| NGC 4350 | A2. 9b | K-2079 | 0.0398* | 1.5, 1.0, 0.5×7 |
| NGC 4501 | A2.10b | K-2145 | 0.0398* | 1.5, 1.0, 0.5×7 |
| NGC 4501 | A2.11b | K-2079 | 0.0398* | 1.5, 1.0, 0.5×7 |
| NGC 4535 | A2.12b | K-2221 | 0.0398* | 1.5, 1.0, 0.5×7 |
| NGC 4535 | A2.13b | K-2090 | 0.0398* | 1.5, 1.0, 0.5×7 |
| NGC 5457 | A2.14d | NS-1872 | 24.0 | 1.5, 1.0, 0.5×7 |
| NGC 4374 | A3. 1b | K-3540 | 24.5 | 1.5, 1.0, 0.5×7 |
| NGC 4429 | A3. 2b | K-2186 | 24.5 | 1.5, 1.0, 0.5×7 |
| NGC 4698 | A3. 3b | K-2220 | 24.5 | 1.5, 1.0, 0.5×7 |
| NGC 3953 | A3. 4b | K-3539 | 24.5 | 1.5, 1.0, 0.5×7 |
| NGC 4535 | A3. 5b | K-2220 | 24.5 | 1.5, 1.0, 0.5×7 |
| NGC 2403 | A3. 6b | NS-917+1847 | 24.0 | 1.5, 1.0, 0.5×7 |
| NGC 5866 | A4. 1b | NS-1886+2364+2371 | 23.5 | 0.5×9 |
| NGC 4762 | A4. 2b | CD-1010[3] | 24.0 | 1.0, 0.5×8 |
| NGC 3115 | A4. 3b | NS-2132+2539 | 22.5 | 1.0×6 |
| NGC 4594 | A4. 4b | NS-2447+2542 | 23.7 | 1.0×6 |
| NGC 4565 | A4. 5b | K-1003+1112+1113 | 25.5 | 1.0, 0.5×5 |
| NGC 4244 | A4. 6b | K-2231 | 25.0 | 1.0×4 |
| NGC 4267 | A5. 2b | K-3540 | 24.5 | 1.5, 1.0, 0.5×8 |
| NGC 4026 | A5. 3b | K-3539 | 24.5 | 1.5, 1.0, 0.5×8 |
| NGC 4374 | A5. 1b | K-3540 | 24.5 | 1.5, 1.0, 0.5×8 |
| NGC 4569 | A5. 4b | K-2730 | 24.5 | 1.5, 1.0, 0.5×8 |
| NGC 4100 | A5. 5b | K-3539 | 24.5 | 1.5, 1.0, 0.5×8 |
| NGC 4234 | A5. 6b | K-3547 | 24.5 | 1.5, 1.0, 0.5×8 |

1) The level of the faintest isophote is given in units of mag arcsec$^{-2}$ or relative intensity (indicated by the asterisk) for calibrated or non-calibrated plates, respectively.
2) From the faintest isophote to brighter ones in units of mag arcsec$^{-2}$.
3) Taken with the du Pont telescope at Las Campanas Observatory by Dr. K. Wakamatsu.